THE PRACTICAL DESIGN OF STRUCTURAL ELEMENTS IN ALUMINIUM

The Practical Design of Structural Elements in Aluminium

John W Bull

AVEBURY
TECHNICAL

Published by
Avebury Technical
Ashgate Publishing Limited
Gower House
Croft Road
Aldershot
Hants GU11 3HR
England

Ashgate Publishing Company
Old Post Road
Brookfield
Vermont 05036
USA

British Library Cataloguing in Publication Data

Bull, John W.
Practical Design of Structural Elements
in Aluminium
I. Title
624.1

ISBN 0 291 39798 0

Typeset in 11 point Baskerville by Photoprint, Torquay, Devon and printed in Great Britain at The University Press, Cambridge.

Contents

List of figures and tables

Preface

The aim of this book is to provide a practical and simplified guide to the design of structural elements in aluminium, using BS 8118 'Structural use of aluminium' as its basis. The book has been specifically written for:

a) those persons in higher education where aluminium design is taught;
b) engineers who wish to design in aluminium, but who are not perhaps as familiar with the structural use of aluminium as they are with such materials as concrete, steel and timber; and
c) those persons who are required to check design calculations.

The book is aimed at building, civil and structural engineers, both at undergraduate level and at postgraduate level, plus consultant and contracting engineers and technicians. Further, aluminium structures are used in the aerospace, offshore and marine industries, where this book will also prove useful.

A book of this size cannot deal comprehensively with all aspects concerning aluminium, its alloys, temper designations and the design of aluminium structures. Consequently, this book can only introduce the subject. Readers who frequently work in the area of aluminium structures will be familiar with *Aluminium Structural Analysis* by P. S. Bulson (Elsevier, London, 1992) and *Aluminum Alloy Structures* by F. M. Mazzolani (Pitman, London, 1985). Both books are detailed texts dealing with the background to the design rules and are appropriate to the more complicated areas of structural aluminium design.

Although this book follows the arrangement of the sections in BS 8118: Pt 1, it has not followed exactly the section divisions. BS 8118: Pt 1 'Code of practice for design' is divided into eight sections and 11 appendices, but this book has placed the 'Appendices' in Chapter 14, the 'British Standards referred to' in Chapter 15, and divided the eight BS sections between the remaining 13 chapters as follows:

Section 1, 'General' of the BS is covered by:
  Definitions and Symbols
Section 2, 'Properties and selection of materials' is covered by:
  Chapter 1 General aspects of structural aluminium
Section 3, 'Design principles' is covered by:
  Chapter 2 Design principles
Section 4, 'Static design of members' is covered by:
  Chapter 3 Static design of members
  Chapter 4 Static design of beams
  Chapter 5 Static design of tension members
  Chapter 6 Static design of compression members
  Chapter 7 Static design of members subjected to bending with axial force and to biaxial bending
  Chapter 8 Static design of members, deformation
Section 5, 'Plates and plate girders' is covered by:
  Chapter 9 An introduction to plates and plate girders
  Chapter 10 Plate girders
Section 6, 'Static design of joints' is covered by:
  Chapter 11 Static design of joints
Section 7, 'Fatigue' is covered by:
  Chapter 12 Fatigue
Section 8, 'Testing' is covered by:
  Chapter 13 Testing

There is also an Introduction before Chapter 1 that indicates those parts of the BS which will be most used in the design of aluminium structures.

Each chapter covers a specific part of structural aluminium design. The chapters start with a précis of the relevant parts, sections and clauses of BS 8118: Pt 1 and, where appropriate, are followed by design examples. The design examples stand on their own and are not part of any single aluminium structure. Each design example concentrates on a specific part of the BS and thus only gives that part of the calculation required for the example. Hence, assumptions are made regarding loading, material dimensions, support conditions, etc., which in a complete set of design calculations would be fully defined.

Within each chapter there is frequent reference to BS 8118: Pt 1 and its sections. The BS sections are called either by the word Section followed by the section number, e.g. Section 5.3.2, or purely by the section number in parentheses as (5.3.2). When using this book, it is necessary for the designer to have to hand BS 8118: Pt 1 for reference to the required figures and tables. Any extracts from BS 8118 are reproduced with the permission of British Standards Institution (BSI). Complete copies of BS 8118 can be obtained from BSI Sales, Linford Wood, Milton Keynes, MK14 6LE: telex 825777 BSIMK G, fax 0908–320856, telephone 0908–220022.

As with all books of a technical nature, errors will creep in no matter how assiduously one may plan otherwise. For any errors in this book I apologise and would thank those persons who take the time to write to me and point out any.

I wish to record my thanks to Dr P. S. Bulson, the Chairman of the BSI Technical Committee, to Mr C. Donahough, Secretary to the BSI Technical Committee and to Mr P. B. Tindall for their guidance. I am also indebted to the Library at DRA, Military Division, RARDE MS6, Christchurch, UK.

Finally, I would like to acknowledge the support of my wife Sonia, whose idea it was I should write this book, and of my son Jonathan, who so often seemed to have a valid reason for using the word processor when I needed it!

Eur Ing Dr John W Bull
BSc, PhD, FIHT, FIWSc, CEng, MIStruct E, MICE

Introduction

Aluminium is light and ductile. Aluminium is about one-third the weight of steel and less susceptible to brittle fracture than steel. A major advantage for aluminium is that any shape of section can be made by extrusion.

Aluminium has a Young's modulus of about one-third that of steel, with a load-deflection curve that is continuous without a yield point. The linear elastic portion of the curve continues to about 0.01 per cent strain. A strain of 0.2 per cent in aluminium corresponds in general terms to the 'elastic limit'; for steel the yield point. The stress at 0.2 per cent strain is denoted as $f_{0.2}$.

Due to its lower Young's modulus, aluminium requires increased design consideration concerning deformation and instability. Bending deformation is important in aluminium beams as is axial deformation in aluminium compression and tension members. In some aluminium truss members, shear deformation can account for up to a third of the total truss deflection. Member buckling is more prevalent in aluminium than in steel and requires increased consideration. Due allowance must be made also for the thermal expansion of aluminium, as it is twice that of steel.

The strength of aluminium may be considered low for structural applications, but it can be increased. For example, cold working will increase the strength of aluminium, but it does reduce ductility. Strength increases can also be obtained by alloying aluminium, but there is a penalty of reduced corrosion resistance. Heat treatment usually increases the aluminium alloy strength.

Welding can be considered as a form of heat treatment! Following welding, residual compression and tension stresses are generated in the heat-affected zone (HAZ). These residual stresses usually reduce the joint strength and the parent material strength, but for some alloys there is no strength reduction. The use of steel bolts will cause contact corrosion unless the bolt is of stainless steel or is in some way protected, e.g. galvanized etc.

Although BS 8118 has now been introduced, it will be some time before CP 118 'The structural use of aluminium' is withdrawn. Many designers who are familiar with CP 118 will wish to develop a professional feel for BS 8118 and to compare a BS 8118 design with the same CP 118 design. Remembering that BS 8118 and CP 118 have a different design philosophy, the major differences likely to be encountered are:

1) more time will be required to read and understand BS 8118;
2) BS 8118 will require increased amounts of calculations and the designs will take longer to finalize;
3) BS 8118 will give higher resistances for beams and will require fewer fasteners and smaller welds; and
4) the advantages of BS 8118 can be more fully exploited if aluminium alloy sections besides those specified in BS 1161 are used.

Many designers who frequently use British Standards (BS 5268, BS 5400, BS 5628, BS 5950 and BS 8110) for structural design understand that the efficient use of a BS can be considerably improved by concentrating on the most frequently used BS sections. In this way, a designer can bring to bear the expertise already gained in designing with other materials. By knowing which areas of aluminium design are most likely to occur repeatedly in design practice, studying them and comparing them with existing knowledge, design errors can be reduced.

The problem in encountering BS 8118: Pt 1 for the first time is that a designer does not know which sections, specific to aluminium design, will eventually most frequently be used. Fortunately, there is a means of determining these sections. Some of the sections in BS 8118: Pt 1 are self-contained and do not refer to other sections, but they may be referred to in other sections. Many sections, however, rely on other sections to provide formulae, figures and tables so that calculations can be made. For example, Section 5.4.5.6 has two formulae, one for a moment calculation and one for a shear calculation. The calculations rely on the use of two tables (4.1 and 4.2), two figures (5.4 and 5.5) and another section (5.4.4.3). If these referrals to other sections are tabulated, it is possible to determine which sections of BS 8118: Pt 1 will probably be most used in the design of aluminium structures.

Table 1 Number of times each section referred to in BS 8118: Pt 1

Section Number	*Section*	*Tables*	*Figures*	*Total*
1	3	–	–	3
2	21	63	–	84
3	19	26	–	45
4	248	86	54	388
5	120	–	29	149
6	31	7	20	58
7	33	19	25	77
8	31	1	–	32
Appendices	74	19	24	117

Table 1 looks at BS 8118: Pt 1 as a whole and gives the number of times each section is referred to within the BS. For example, Section 4 and its accompanying sub-sections are referred to 248 times, with its tables being referred to 86 times and its figures, 54 times. Clearly:

Section 4 'Static design of members' would appear to be, and in fact will be, the most used section, with
Section 5 'Plates and plate girders' being the next most used section.

This means that in any aluminium design, persistent use will be made of these sections. What must also be remembered is that for every aluminium design,

Section 2 'Properties and selection of materials' and
Section 3 'Design principles'

will always be used, although this use will be restricted to the initial stages of the design.

Moving on to the level of sub-sections in BS 8118: Pt 1, Table 2 shows how often each sub-section is referred to. Table 2 shows the sub-sections most likely to be encountered when designing structural aluminium, that is:

Table 2 Sub-sections most frequently referred to in BS 8118: Pt 1

Section Number	*Sub-section Number* .1	.2	.3	.4	.5	.6	.7	.8	.9	.10	
1	0	1	1	–	–	–	–	–	–	–	
2	0	7	2	3	6	0	1	–	–	–	
3	0	2	1	2	1	7	1	0	–	–	
4	0	3	37	46	62	14	44	26	2	–	
5	0	21	19	71	–	–	–	–	–	–	
6	0	0	0	10	5	0	1	1	8	0	
7	0	0	5	4	5	3	3	5	–	–	
8	0	0	4	21	1	–	–	–	–	–	
Appendices	A	B	C	D	E	F	G	H	J	K	L
	2	5	2	4	3	6	2	6	3	4	2

a) 5.4 'Plate girders',
b) 4.5 'Beams',
c) 4.4 'HAZ softening adjacent to welds',
d) 4.7 'Compression members', and
e) 4.3 'Section classification and local buckling'.

For the Appendices, shown in Table 2, the number underneath the Appendix letter shows how often each appendix is referred to in the main body of BS 8118: Pt 1. An experienced designer would expect to find:

Appendix F 'HAZ adjacent to welds' and
Appendix H 'Lateral torsional buckling of beams'

to be prevalent, as both areas are of considerable significance in the design of aluminium structures.

The final table, Table 3, shows the figures and the tables most often referred to in BS 8118: Pt 1. As a designer would expect,

the limiting stresses (Tables 4.1 and 4.2),
the material factors (Table 3.3) and
the alloys (Tables 2.1, 2.2 and 2.3)

are referred to most often.

Welding is well represented (Tables 4.5, 7.1, 7.2 and 7.3) as is section classification (Table 4.3).

The figures referred to most often relate to:

Table 3 Figures and tables referred to more than five times in BS 8118: Pt 1

Number of times referred to	*Figures*	*Tables*
26–30		4.1
21–25		3.3, 4.2
16–20		2.1, 2.2
11–15	4.10	4.3
6–10	4.2, 4.5, 4.6, 4.12, 5.2, 6.3	2.3, 4.5, 7.1, 7.2, 7.3

buckling (Figures 4.2, 4.5, 4.10, and 4.12) and
heat-affected zones (Figures 4.6 and 6.3).

Please remember that the BS 8118 assumes an appropriately qualified and experienced person will be available to carry out a design, and compliance with the BS does not give immunity from legal obligations.

Definitions (1.2)

The definitions used in this book are the same as those given in Section 1.2 of BS 8118: Pt 1. However, BS 8118: Pt 1 includes additional definitions not given in Section 1.2 and these are included below.

Action Effect under Factored Loading (S_d): The effect of the factored load or combination of factored loads on the structural member.
Characteristic Action (F_k): The load to which a structure may be expected to be subjected during normal service. The nominal load has an expected return period of not less than the structure's specified design life.
Characteristic Resistance (R_k): Is a modified mean resistance R_m, equal to $R_m(1-kv)$.
Compact Cross-Section: A cross-section that can develop the full plastic capacity, in either bending or compression, with no reduction due to local buckling of thin-walled elements.
Damage-Tolerant Design: Where the damage to a structure under fatigue conditions is assessed by inspection monitoring of the rate of growth of fatigue cracks.
Dead Load: Structure self-weight and any permanently attached items.
Design Action (F_d): The nominal load (F_k) multiplied by the relevant overall load factor (γ_f).
Design Action Effect (S_d): The effect of the factored load or combination of factored loads on the structural member.
Design Life: The period in which the structure or component is

required to perform safely, with an acceptable probability that it will not require repair or withdrawal from service. Typical values of design life are 120 years for a highway bridge to ten years for a vehicle body.
Design Resistance (R_d): The resistance of a member divided by the relevant partial material factor = (R_k/γ_m).
Design Load Spectrum: A tabulation of the numbers of occurrences of all the stress ranges caused by loading events.
Design Spectrum: A tabulation of the numbers of occurrences of all the stress ranges caused by loading events.
Design Stress Range: The maximum fatigue load stress minus the minimum fatigue load stress, divided by the fatigue material factor γ_{mf}.
Detail Class: A rating given to a detail which indicates its level of fatigue resistance.
Edge Distance: The distance from the centre of a fastener hole to the nearest edge of an element.
Effective Length: The length between points of effective restraint of a member, multiplied by a factor to take account of end conditions and loading.
Effective Section:
Includes:
a) the reduced thickness in the vicinity of welds to allow for HAZ softening and local buckling but
b) with no deduction for holes.

Estimated Life: The factored design life/(Σ n/N)
Factored Design Life:
a) The estimated life times (Σ n/N).
b) The nominal design life times the fatigue life factor γ_L.
c) Is $\leq$ (N_a/F).

Factored Life: The design life multiplied by the relevant partial life factor.
Factored Load (F_d): The nominal load (F_k) multiplied by the relevant overall load factor (γ_f). The factored load is used for checking the limit state of static strength.
Factored Resistance (R_d): The calculated resistance (R_k) of a member divided by the relevant partial material factor γ_m.
Fail Safe: The ability of a structure to continue to be serviceable after the discovery and monitoring of fatigue cracks.
Fatigue: The damage, by gradual cracking, to a structural member caused by repeated applications of a stress that is insufficient to cause failure by a single application. The fatigue design load spectrum uses the unfactored loads.
Fusion Boundary: The material in a heat-affected zone immediately adjacent to the leg of a weld.

Heat-Affected Zone (HAZ): A zone in which there is a reduction in strength of the material in the vicinity of welds in certain classes of alloy.
Imposed Loads: All statically or dynamically applied loads on a structure other than dead or wind loads. For combinations of imposed loads, the imposed load is $\gamma_{f2} \times$ nominal loads.
Instability: A loss or often sudden loss of stiffness of a structure, that limits its load-carrying capability and in certain instances can cause catastrophic failure.
Lateral Torsional Buckling: The buckling of a beam accompanied by a combination of lateral displacement and twisting.
Lateral Restraint: Restraint that limits lateral movement of the compression flange of a beam.
Life to Failure: Determined by test is (N_a/F) and equals the log mean life to failure divided by a fatigue test factor F.
Limit State: Condition beyond which a structure is unfit for its intended use.
Loading Event: A defined loading cycle that for design purposes is assumed to repeat a given number of times.
Load Spectrum: A tabulation of the relative frequencies of loading events of different intensities on a structure.
Local Buckling: Buckling of the thin walls of a component in compression, characterized by the formation of ripples or waves along the member.
Miner's Summation: A cumulative fatigue damage summation based on a rule devised by Palmagren and Miner.
Net Effective Section:
Includes:
a) the deduction for holes and
b) the reduced thickness in the vicinity of welds to allow for HAZ softening.

Net Section: Includes the deduction only for holes.
Nominal Load (F_k): The load to which a structure may be expected to be subjected during normal service. The nominal load has an expected return period of not less than the structure's specified design life. Nominal loads are used for checking the limit states of deformation, fatigue and vibration.
Nominal Resistance (R_k): Is a modified mean resistance R_m equal to $R_m(1-kv)$.
Outstand Element: The element of a section, composed of flat or curved elements, which is supported along one longitudinal edge and free along the other.
Reinforced Element: The element of a section that is stiffened by the

introduction of longitudinal reinforcement, either along the edge of the element or within its width.

Resistance: The strength of a member based on calculations, using acceptable maximum values for material strength.

Safe Life: A design against fatigue in which the calculated life is many times longer than the life required in service.

Semi-Compact Cross-Section: A cross-section of a beam in which the stress in the extreme fibres is limited to the 0.2 per cent proof stress, because local buckling of the compression elements would prevent development of the full plastic moment capacity.

Serviceability Limit State: Those limit states that when exceeded can lead to the structure being unfit for its intended use, even though the structure has not collapsed.

Slenderness: The effective length of a strut divided by the radius of gyration.

Stiffened Elements: The element of a section, composed of flat or curved elements, which is supported along both longitudinal edges.

Stress Cycle: A pattern of variation of stress at a point, which is normally in the form of two opposing half-waves.

Stress History: A record showing how the stress at a point varies during loading.

Stress Range: The greatest algebraic difference between the principal stresses occurring on the principal planes not more than 45° apart in any stress cycle on a plate or element.

Stress Range: The algebraic or vector difference between the greatest and the least vector sum of stresses in any one stress cycle on a weld.

Stress Spectrum: A tabulation of the numbers of occurrences of all the stress ranges of different magnitudes during a loading event.

Structural Adequacy: Where $R_d > S_d \gamma_c$.

Temperature Effects: Forces developed due to temperature changes.

Torsional Buckling: Buckling of a strut accompanied by twisting.

Torsional/Flexural Buckling: Buckling of a strut accompanied by overall flexure as well as twisting.

Ultimate Limit State: Those limit states that when exceeded can cause collapse of part or the whole of a structure.

Unfactored Loads: The load to which a structure may be expected to be subjected during normal service. The nominal load has an expected return period of not less than the structure's specified design life. Unfactored loads are used when checking fatigue.

Wind Loads: Dynamic loads due to wind gusts. Wind loads can be the most severe load applied to the structure.

Symbols (1.3)

The symbols used in this book are the same as those used in BS 8118. Section 1.3 of BS 8118: Pt 1 defines the major symbols, but in Sections 2 to 8 inclusive, additional definitions and alternative definitions are used. For this reason, this book has included an extensive list of all the symbols used in BS 8118: Pt 1. Some symbols are used exclusively in BS 8118: Pt 2 and are not necessarily included in the following list.

The symbols in the following table are listed alphabetically, with upper case letters preceding lower case letters, followed by upper case Greek letters and lower case Greek letters. Within each section of four letter sections, the subscript order is alphabetical, irrespective of case. Numbers follow in numerical order.

Symbol		*Definition*
A	=	Area
A	=	Durability rating
A	=	Gross area of the entire cross-section of the plating
A	=	Gross section area
A	=	Gross section area with no reduction for HAZ softening, holes or local buckling
A	=	Percentage area (of porosity) as seen on a radiograph in any 100 mm of weld length
A	=	Section area

A	=	Total area of all the weld deposits
A_e	=	Effective section area
A_e	=	Section area
A_{es}	=	A_h
A_{es}	=	A_{SH}, the area of the shank, when the shear plane passes through the shank
A_{es}	=	A_{tb}
$A_e\bar{y}$	=	First moment of the gross excluded area outside the web
A_h	=	Hole area
A_h	=	Hole area with: $K_1 = 1.0$ for rivets, $K_1 = 0.95$ for close tolerance bolts and $K_1 = 0.85$ for normal clearance bolts
A_h	=	Sum of the hole areas on the diagonal or zig-zag section
A_n	=	Net section area, with a deduction for holes
A_n	=	Net section area, with a deduction for unfilled holes
A_{ne}	=	Effective area obtained by taking a reduced thickness to allow for buckling as well as HAZ softening, but with holes ignored
A_{ne}	=	Net effective section area
A_{ne}	=	Net effective section area, with a deduction for holes
A_o	=	Effective area of any outstanding legs lying clear of the connected element, but ignoring any fillet
A_{SH}	=	Area of the shank, when the shear plane passes through the shank
A_{tb}	=	Stressed area of the threaded part of the bolt, when the shear plane passes through that area
A_{te}	=	Effective section area of the tongue plate, or the total area for two tongue plates
A_v	=	Effective shear area
A_{we}	=	Effective section area of the web plate between the flanges, or to the tongue plate tips
a	=	B, the width of plating
a	=	Spacing of the main components measured to the centroids of the connections to each batten
a	=	Spacing of the transverse stiffeners
a	=	Width of an unstiffened plate
a	=	0.6 for aluminium bolts
a	=	1.0 for stainless steel and steel, bolts and rivets
a_o	=	The distance from the stiffener to the free edge of the web plate

B	=	Durability rating
B	=	Overall width of a multi-stiffened plate
B	=	Total flange width
B	=	Width of the plating
B_{RF}	=	Single fastener factored resistance in bearing
B_{RP}	=	Bearing capacity of the connecting ply
b	=	Developed width of an element at mid-metal
b	=	Width of flat elements
b	=	Width of an element generally
b_e	=	Effective width of a web plate (plate girder)
b_1	=	Distance either side of a stiffener attachment/s
C	=	Durability rating
C	=	Maximum overall as measured depth
c	=	Clear depth of the rib or lip measured to the surface of the plate
c	=	1.5 when $(d_f/t) > 13$
c	=	$(20t/d_f)$ when $10 < (d_f/t) < 13$
c	=	2 when $(d_f/t) < 10$
D	=	As measured average depth
D	=	As specified on drawings
D	=	Diameter of round tube, to mid-metal
D	=	Overall depth of the web to the outside of the flanges
D	=	Overall section depth
D	=	Total depth
DF	=	Applies only where weld caps have to be dressed flush
DS	=	Dress smoothly
D_1	=	Internal diagonal length of the box
D_2	=	Internal diagonal length of the box
d	=	Clear depth of a web between the flanges, which for inclined webs is measured on the slope
d	=	Depth of the largest sub-panel
d	=	Depth of the unstiffened plate
d	=	Depth of the web between flanges
d	=	Depth of the web element in a beam
d	=	Depth of the web measured between the flanges or to the tongue plate tips
d	=	Least panel dimension
d	=	Longitudinal spacing between the centres of the battens

d	=	Maximum distance of a point in the weld group from the centroid
d	=	$0.61([wt^3]/[I_{SU}])^{0.375}$
d_{av}	=	Average depth of the two sub-panels lying either side of the longitudinal stiffener
d_f	=	Nominal diameter of the fastener
d_f	=	Nominal pin diameter
d_z	=	Total depth of HAZ material within the clear depth of the web, between the flanges
E	=	Modulus of elasticity
E	=	Refer to designer
e	=	Distance from the centre of the hole to the adjacent edge, in the direction the fastener bears
e	=	Maximum off-centre of the web
F	=	Factor dependent upon the effective number of test results available
F	=	Fatigue test factor
F	=	Fusion boundary of the heat-affected zone (HAZ)
F	=	Maximum out-of-flatness of the flange
Fat	=	Fatigue number
F′	=	Maximum out-of-flatness of the flange
F_c	=	Friction capacity of a high strength friction grip (HSFG) bolt
F_d	=	Factored load or design action = $\gamma_f F_k$
F_k	=	Nominal load or characteristic action
f	=	Reduction factor applied to k_z
f_r	=	Amplitude of the fatigue cycles
f_r	=	Design stress range
f_r	=	Principal stress range at the detail and is constant for all cycles
f_{r1}	=	Rank of the cycles in descending order of amplitude (f_{r1}, f_{r2}, etc.)
f_{max}	=	Maximum fatigue load stress
f_{min}	=	Minimum fatigue load stress
f_{oc}	=	Constant amplitude cut-off stress
f_{ov}	=	Variable amplitude cut-off stress
f_u	=	Guaranteed minimum tensile strength of the material
f_u	=	Ultimate tensile stress (Rm in BS EN 10002–1)
$f_{0.2}$	=	Guaranteed minimum 0.2 per cent parent metal proof

		stress, if it is in the annealed O condition
$f_{0.2}$	=	Guaranteed minimum tensile 0.2 per cent proof stress
$f_{0.2}$	=	Minimum 0.2 per cent tensile proof stress ($R_{p0.2}$ in BS EN 10002–1)
$f_{0.2}$	=	Minimum 0.2 per cent tensile proof stress of the plate material
G	=	Shear modulus
g	=	Stress gradient coefficient
g_l	=	Leg length of a weld
g_t	=	Throat of weld
g_t	=	Throat width of a fillet weld
H	=	Warping factor
HAZ	=	Heat-affected zone
h	=	Distance to a free edge
h	=	Reinforced element's coefficient
I	=	Second moment of area of the gross cross-section
I_{ep}	=	Second moment of area of the end-post section about the centre-line of the web
I_o	=	Polar second moment of area about the centroid of the weld group
I_S	=	Second moment of area of the full section of the effective stiffener (plate girder)
I_S	=	Second moment of area of the full section of the effective stiffener about a centroidal axis parallel to the web
I_{SU}	=	Second moment of area of one sub-unit of the plating (multi-stiffened plates)
I_{SU}	=	Second moment of area of one sub-unit of the plating as defined in Section 5.3.2.3, about a centroidal axis parallel to the plane of the plating
I_y	=	Second moment of inertia about the centroid axis
I_y	=	Second moment of inertia about the minor centroid axis
J	=	Torsional constant
K	=	Effective length factor for struts
K_1	=	Coefficient used in the calculation of resistance in bolts
K_1	=	0.85 for normal clearance bolts
K_1	=	0.95 for close tolerance bolts

K_1 = 1.0 for rivets
K_2 = Constant depending upon the class of detail to ensure a high probability of survival
K_2 = Constant used in fatigue failure criterion

k = Coefficient depending on the probability distribution function
k = Strut torsional buckling interaction factor found from Figure 4.11
k_L = Effective thickness factor for web buckling
k_L = Local buckling coefficient
k_v = Reduction factor on longitudinal resistance to take account of high shear
k_v = $1.6 - 1.2(V/V_{RS})$
k_z = Softening factor for the HAZ material
k_z = Strength factor for the HAZ material
k'_z = Modified strength factor for the HAZ material
k'_z = Modified softening factor found from Appendix F.2
k_{zy} = $k_z + 1 - (y/y_1)$

L = Distance between restraints
L = Length between points of lateral supports
L = Length between supports
L = Length of a cantilever strut
L = Straight joint length
L = Straight weld joint length
L = Total weld length

l = Effective buckling length
l = Effective length
l = Effective length between lateral supports
l = Effective length for lateral torsional buckling
l = Effective length of the plating
l = Effective strut length
l = Length parallel to weld axis
l_e = Effective length of the butt weld
l_f = Effective length of the fillet weld

M = Equivalent uniform moment
M = In-plane moment due to the factored loading
M = Moment under factored loading
M = Moment under factored loading (in the length being considered)

MIG	=	Metal inert-gas welding
$\bar{M}$	=	Equivalent uniform moment
M_{cr}	=	Elastic critical uniform moment for lateral torsional buckling
M_{ep}	=	Factored loading moment acting in the plane of the web, to be resisted by the end-post
M_f	=	Fully compact value of M_{RS}
M_{RF}	=	Reduced value of M_{RS} for the flanges on their own, with the web omitted, i.e. for the flanges only
M_{RPx}	=	Lower value of $\bar{M}_x$ that would be acceptable in combination with P, but in the absence of minor axis bending as given in Section 4.8.4.2
M_{RPy}	=	Value of $\bar{M}_y$ that would be acceptable in combination with P, but in the absence of major axis bending as given in Section 4.8.4.4
M_{RS}	=	Factored in-plane moment resistance of a section, reduced to allow for coincident shear if necessary (5.2.6.4)
M_{RS}	=	Factored moment resistance of a section in the absence of shear
M_{RS}	=	Factored resistance to in-plane moment reduced to allow for any coincident high shear (5.2.6.4)
M_{RSO}	=	Reduced factored moment resistance of a section to allow for shear
M_{RSx}	=	Factored uniaxial moment resistance about a major axis, with an allowance for shear
M_{RSx}	=	Factored uniaxial moment resistance of the cross-section, about a major axis, adjusted to allow for any coincident shear
M_{RSy}	=	Factored uniaxial moment resistance about a minor axis, with an allowance for shear
M_{RSy}	=	Factored uniaxial moment resistance of the cross-section, about a minor axis, adjusted to allow for any coincident shear
M_{Rx}	=	$p_s S/\gamma_m$
M_{Rx}	=	Factored moment resistance to lateral torsional buckling
M_s	=	Semi-compact value of M_{RS}
M_u	=	Ultimate moment
$\bar{M}_{uf}$	=	Sp_o
M_{us}	=	Moment corresponding to the adopted stress pattern
M_x	=	Equivalent uniform moment about the major axis
M_x	=	Factored uniaxial bending moment about a major axis
M_x	=	Uniaxial moment about a major axis

$\bar{M}_x$	=	Equivalent uniform moment about the major axis as obtained in Section 4.5.6.4
M_y	=	Equivalent uniform moment about the minor axis
M_y	=	Factored uniaxial bending moment about a minor axis
M_y	=	Uniaxial moment about a minor axis
$\bar{M}_y$	=	Equivalent uniform moment about the minor axis as obtained in Section 4.5.6.4
M_1	=	Maximum factored moment
M_2	=	Minimum factored moment
m	=	Inverse slope of the f_r – N curve (fatigue) and is constant for most detail classes
m_1	=	Shear buckling factor found from Figure 5.7 for plate girders
m_2	=	$([4p_{of}S_f]/[p_{ow}d^2t])^{0.5}$, a shear buckling factor for plate girders
N	=	Endurance in cycles and equals (K_o/f_r^m), the mean line; or equals (K_2/f_r^m), the design line
N	=	Number of battens at each position
N	=	Number of webs
N	=	Predicted number of cycles to failure of a stress range f_r
NA	=	Not applicable
NA	=	Not normally applicable
NL	=	No limit
NP	=	Not permitted
N_a	=	Log mean life to failure
N_F	=	Number of friction interfaces
n	=	Frequency of the fatigue cycles
n	=	Time in days between welding and loading
n	=	Equivalent number of cycles in the stress range (fatigue)
P	=	Axial compressive force due to factored loading
P	=	Axial factored load
P	=	Axial force due to factored loading
P	=	Axial tensile force due to factored loading
P	=	Protection
P	=	Uniform in-plane compression
PP	=	Partial penetration
P_{cr}	=	Elastic critical load for torsional buckling
P_{cr}	=	Elastic critical load for torsional buckling, allowing for interaction with column buckling when necessary
P_o	=	Proof load for a bolt

P_p	=	Pre-stress load
P_p	=	Pre-stress load for a HSFG bolt
P_R	=	Factored axial resistance based on the overall column or torsional buckling
P_{RB}	=	Factored resistance of a butt weld
P_{RF}	=	Factored resistance of a fillet weld
P_{RFB}	=	Factored direct normal tensile force resistance of a HAZ adjacent to a butt weld fusion boundary
P_{RFF}	=	Factored direct normal tensile force resistance of a HAZ adjacent to a fillet weld fusion boundary
P_{RG}	=	Factored resistance of a bonded joint
P_{RS}	=	Factored axial compressive resistance
P_{RS}	=	Factored axial resistance of the section (4.6.2 or 4.7.7)
P_{RS}	=	Factored axial resistance, reduced to allow for any coincident high shear (5.2.6.4)
P_{RS}	=	Factored resistance in axial tension
P_{RT}	=	Factored resistance for a single fastener in axial tension
P_{RTB}	=	Factored direct normal tensile force resistance of a HAZ adjacent to a butt weld toe
P_{RTF}	=	Factored direct normal tensile force resistance of a HAZ adjacent to a fillet weld toe
P_{Rx}	=	Factored axial resistance to overall column buckling about the major axis (4.7.3)
P_{Ry}	=	Factored axial resistance to overall column buckling about the minor axis (4.7.4)
P_{RZ}	=	Factored resistance of the HAZ under direct loading
P_l	=	Concentrated load or reaction acting at the stiffener

p	=	p_m for a slender element
p	=	p_o for a fully compact or semi-compact element
p_a	=	Lesser of $0.5(f_{0.2} + f_u)$ and $1.2f_{0.2}$ for the material of the connected ply
p_a	=	Limiting stress for local capacity (compression and tension)
p_a	=	Limiting stress for parent metal as given in Table 4.1 or by D.1
p_a	=	Limiting stress for parent metal in the condition used
p_{az}	=	Limiting direct stress in the HAZ
p_{a0}	=	Limiting stress for parent metal in the 0 condition
p_{a6}	=	Limiting stress for parent metal in the T6 condition
p_f	=	Guaranteed minimum yield stress for the bolt or rivet stock
p_f	=	Lesser of $0.5(f_{0.2} + f_u)$ and $1.2f_{0.2}$
p_f	=	Limiting stress for a solid bolt or a solid rivet
p_m	=	Limiting compressive stress in a slender element

p_o	=	Limiting stress for bending and overall yielding
p_o	=	Limiting stress for the material
p_o	=	Limiting stress for parent metal in the condition used
p_{of}	=	Limiting stress for the flange material
p_{ow}	=	Limiting stress for the web material
p_{o0}	=	Limiting stress for parent metal in the O condition
p_{o6}	=	Limiting stress for parent metal in the T6 condition
p_s	=	Buckling stress for a column sub-unit
p_s	=	Buckling stress for web treated as a thin column between flanges
p_s	=	Column buckling stress
p_s	=	Lateral torsional buckling stress
p_s	=	Limiting stress for overall buckling stability
p_s	=	Overall buckling stress
p_t	=	Weld penetration
p_v	=	Limiting stress in shear
p_v	=	Limiting stress in shear for the web material
p_{vt}	=	Limiting stress for the tongue plate materials
p_{vw}	=	Limiting stress for the web materials
p_{vz}	=	Limiting shear stress for the material in the HAZ
p_w	=	Limiting stress for weld metal
p_{w1}	=	Stress arising at the extreme edge, assuming a 45° angle of dispersion either side of the localized force
p_{w1}	=	Stress arising at the extreme edge of a web due to a localized force
p_{w2}	=	The stress arising at the mid-point, assuming a 45° angle of dispersion either side of the localized force
p_{w2}	=	The stress arising at the mid-point of a web due to a localized force
p_1	=	Stress axis value of p_s in the strut curve diagrams
p_1	=	Value of p_o for unwelded fully compact sections
p_1	=	$([\gamma_m M_{RS}]/S)$ for all sections including hybrid sections, except for unwelded fully compact sections.
q	=	Mean shear stress arising in the end-panel of the web due to factored loading, based on the unreduced thickness
R	=	Radius of curvature of a curved internal element, to mid-metal
R	=	Reaction at the end of the girder, under factored loading
R	=	Repair by welding to approved procedure
R_d	=	Design resistance or factored resistance = R_k/γ_m

R_d = Design resistance or factored resistance $> S_d/\gamma_c$ for structural adequacy
R_f = Ratio of flange area at the point of minimum moment to that at the point of maximum moment between adjacent restraint points, and refers to the ratio of the total area of both flanges or to the area of the compression flange only, whichever gives the lower value
R_k = Characteristic or nominal resistance $= R_m(1-kv)$
R_m = Mean resistance of the failing loads

r = Radius of gyration
r_y = Minor axis radius of gyration of the section.

S = Direction of the external loading action
S = Plastic section modulus
S = Plastic section modulus of the gross section, with no reduction for HAZ, holes or local buckling
SS = Single-sided
S/S = Within 20 mm of a stop or start
S_a = External loading actions under factored loading
S_a = External loading actions under factored loading on the HAZ, in direct loading
S_b = External loading actions under factored loading
S_b = External loading actions under factored loading on the HAZ, in shear
S_d = Action effects under factored loading
S_d = Design action effects
S_d = Standard deviation of the failing loads
S_f = Plastic modulus of the effective flange section (plate girder)
S_f = Plastic modulus of the effective flange section about its own equal area axis, in the plane of the web. The lower of the flange values is used if the flanges are different
S_n = Plastic modulus of the net section
S_{ne} = Plastic modulus of the net effective section
S_{tb} = Applied external tensile load in the axial direction of the bolt

s = Factor on p_l to allow for a strut not meeting the tolerances of straightness or twist
s = Factor to allow for a strut not meeting the tolerances of straightness or twist
s = Torsional buckling parameter for struts, found from Table 4.9

T	=	Toe of the HAZ
TIG	=	Tungsten inert-gas welding
T_o	=	Interpass temperature
t	=	General plate thickness
t	=	Plate thickness
t	=	Thickness
t	=	Thickness of the connected ply
t	=	Thickness of the thinnest outside ply
t	=	True member thickness
t	=	Unreduced web plate thickness
t	=	Web thickness
t_A	=	Lesser of $0.5(t_B + t_C)$ and $1.5t_B$
t_B	=	Thickness of the thinnest element connected by the welding
t_C	=	Thickness of the thickest element connected by the welding
t_e	=	Effective throat thickness of the weld
t_f	=	Flange thickness
t_f	=	Flange thickness, taken as the maximum value if the thickness varies along the girder
t_2	=	Flange thickness
V	=	Average value of the shear force arising in the web panels either side of the stiffener considered
V	=	Shear force under factored loading
V	=	Shear load under factored loading
V_{ep}	=	Factored shear force loading to be resisted by the end-post
V_{RFB}	=	Factored shear resistance of a HAZ adjacent to a butt weld fusion boundary
V_{RFF}	=	Factored shear resistance of a HAZ adjacent to a fillet weld fusion boundary
V_{RS}	=	Factored resistance in shear
V_{RS}	=	Factored resistance of a single bolt or rivet in single shear
V_{RS}	=	Factored shear force resistance
V_{RTB}	=	Factored shear resistance of a HAZ adjacent to a butt weld toe
V_{RTF}	=	Factored shear resistance of a HAZ adjacent to a fillet weld toe
V_{RW}	=	Reduced value for V_{RS} found by putting $m = 0$
V_{RZ}	=	Factored resistance of the HAZ in shear
v	=	Coefficient of variation of the distribution function
v	=	Relative mean quadratic deviation of the distribution of material strength found by tests

v_{tf} = Tension field factor
v_{tf} = Tension field factor for plate girders
v_1 = Elastic critical shear buckling factor
v_1 = Initial shear buckling factor read from Figure 5.4
v_2 = Basic tension field shear buckling factor
v_2 = Shear buckling factor found from Figure 5.5
v_3 = Flange-assisted tension field buckling factor
v_3 = Shear buckling factor found from Figure 5.6

W = Total factored loading on the adjacent span
W = Weld metal
WE = Within 20 mm of the weld end

w = Pitch of the corrugations or stiffeners in a multi-stiffened plate

X = Torsional buckling parameter for struts, found from Table 4.9.
X = 1.0 or the value found from Table 4.7
X_A = Distance of the point A from the centre-line of a butt weld or the root of a fillet weld
X_B = Distance of the point A from the centre-line of a butt weld or the root of a fillet weld

x = Longitudinal pitch of the holes
x = 0.4 times the effective buckling length l

Y = Torsional buckling parameter for struts, found from Table 4.9.
Y = 0.05 or the value found from Table 4.7

y = Distance from the centre of a multi-stiffened plate to the centre of the outermost stiffener
y = Transverse pitch of the holes
$\bar{y}$ = Distance of the centroid of the neutral axis to the section area
y_c = Distance from the neutral axis of the gross section to the more heavily compressed edge, taken as positive towards the compression edge
y_c = Distance from the neutral axis to the more heavily compressed edge
y_o = Distance from the neutral axis of the gross section to the compression flange element in a beam

y_o	=	Distance from the neutral axis of the gross section to the less heavily compressed edge or edge in tension, taken as positive towards the compression edge
y_o	=	Distance from the neutral axis to the less heavily compressed edge or edge in tension
y_1	=	Distance from the buckling axis to the further extreme fibre
y_1	=	Distance from the elastic neutral axis to the extreme fibres
y_1	=	Distance from the gross neutral axis to the compression edge
y_1	=	Distance from the neutral axis to the most severely stressed fibres in a beam
y_2	=	Distance from the buckling axis to the nearer extreme fibre
y_2	=	Distance from the elastic neutral axis of the gross section to the element being considered
y_2	=	Distance from the neutral axis to the compression flange element in a beam
Z	=	Elastic section modulus of the full cross-section of the plating for in-plane bending
Z_e	=	Elastic modulus of the effective section
Z_n	=	Elastic modulus of the net section
Z_{ne}	=	Elastic modulus of the net effective section
z	=	Distance the HAZ extends from a weld
z_o	=	Basic value of z, the distance the HAZ extends from the weld
α	=	Ratio of minimum to maximum shear stress in the web assuming an elastic stress distribution
α	=	Modifying factor for the extent of the HAZ to allow for elevated temperature
α_S	=	Coefficient used in the calculation for bolts or rivets in single shear
α_s	=	0.6 for aluminium bolts and rivets and
α_s	=	0.7 for steel bolts or rivets
β	=	Slenderness parameter
β	=	Value of β for the most critical element in the section
β_o	=	Semi-compact limiting value of β
β_1	=	Fully compact limiting value of β
γ_c	=	Factor for the consequence of failure
γ_f	=	Load factor

γ_f	=	Overall load factor
γ_{f1}	=	Partial load factor for load type
γ_{f2}	=	Partial load factor for load combination
γ_L	=	Fatigue life factor
γ_m	=	Material factor
γ_m	=	Material factor, equal to 1.2 for all aluminium, stainless steel and steel bolts and rivets
γ_m	=	Material factor, equal to 3 for bonded joints
γ_{mf}	=	Fatigue material factor
γ_s	=	Coefficient in the calculation of friction capacity
γ_s	=	1.1 if the value of μ_s is found from tests
γ_s	=	1.33 if the value of μ_s is taken as 0.33
Δ_D	=	Deviation from verticality of the web at a support
$\Delta_{F'}$	=	Maximum out-of-flatness of the flange
Δ_W	=	Maximum distortion of the web
δ	=	Factor used to modify z
δ	=	Modifying factor related to the distance the HAZ extends from the weld
δ	=	1 or 0.75 or $(1.5 + [1.3L/z_o])/(3 + [L/z_o])$
ϵ	=	Constant $(150/p_o)^{0.5}$
ϵ	=	Constant $(250/p_y)^{0.5}$
η	=	Modifying factor for the extent of the HAZ to allow for increased heat build-up
θ	=	Angle between the line of the butt weld and the line of action of the external load
θ	=	Angle between the plane of the applied loading and the web
λ	=	Slenderness parameter for column buckling, lateral torsional buckling and torsional buckling
λ_o	=	Slenderness parameter for column buckling, lateral torsional buckling and torsional buckling
λ_t	=	Torsional buckling parameter for struts, found from Table 4.9
λ_u	=	Effective slenderness for column buckling about the u–u axis as defined in Table 4.9
λ_x	=	Effective slenderness for column buckling about the x–x axis as defined in Table 4.9

λ_y	=	Slenderness ratio of a strut about the minor (y–y) axis
λ_1	=	Slenderness parameter for column buckling of the complete member about an axis parallel to the battens
λ_2	=	Slenderness parameter for column buckling of the complete member about an axis normal to the battens
λ_3	=	Slenderness parameter for the buckling of one of the main components between the battens, based on the most critical of either column or torsional buckling.
λ_4	=	Slenderness parameter for column buckling, lateral torsional buckling and torsional buckling
λ_8	=	Slenderness parameter for column buckling, lateral torsional buckling and torsional buckling
μ_s	=	Slip coefficient
μ_s	=	Slip factor
μ_s	=	Slip factor, usually equal to 0.33
ρ	=	Factor depending on the amount of fillet material at the root of the section (Table 4.9)
ΣA_h	=	Sum of the hole areas on the diagonal or zig-zag section being considered
σ_{cr}	=	Elastic critical stress of an element with reinforcement
σ_{cro}	=	Elastic critical stress of an element without reinforcement
σ_1	=	Normal stress on the weld under factored loading
σ_1	=	Normal stress perpendicular to the throat section under factored loading
τ_1	=	Shear stress acting on the throat section perpendicular to the axis of the weld under factored loading
τ_1	=	Shear stress perpendicular to the weld axis
τ_2	=	Shear stress acting on the throat section parallel to the axis of the weld under factored loading
τ_2	=	Shear stress parallel to the weld axis

1 General aspects of structural aluminium

1.1 Introduction

BS 8118, 'Structural use of aluminium', is in two parts: Part 1, 'Code of practice for design' and Part 2, 'Specification for materials, workmanship and protection'. BS 8118 has a different design philosophy to CP 118, 'The structural use of aluminium'. BS 8118 is based on limit state principles and incorporates much of the latest European research. At the present time both BS 8118 and CP 118 can be used. A Eurocode on the structural use of aluminium is projected.

Part 1 is a code of practice for the design and testing of aluminium structures. It considers the design of elements of framed, lattice and stiffened plate structures, using wrought aluminium alloy. The design recommendations are for aluminium alloys suitable for such structural purposes as bridges, buildings, cranes, marine craft, offshore topside structures, rail and road vehicles, and towers, all subjected to normal atmospheric conditions. Castings and forgings must be designed and manufactured in accordance with the appropriate British Standard and in consultation with the manufacturer (1.1).

Part 2 gives the requirements for the materials, protection and workmanship which apply to assembly, erection, fabrication and protection of aluminium and aluminium alloy construction and structures.

BS 8118 assumes that the design of aluminium structures will be entrusted to appropriately experienced and qualified persons, with construction and supervision being carried out by capable and experienced organizations.

1.2 Materials (2.1)

BS 8118 classifies wrought aluminium and aluminium alloys used for general engineering purposes in accordance with the international four-digit system. Generally, BS 8118 uses the alternative alloy temper designation of ISO 2107. Castings are designated in accordance with BS 1490.

1.2.1 Alloy groups (Appendix A)

The alloy designation uses four digits as follows:

1. Aluminium 99.00 per cent minimum and greater,	1***
2. Aluminium alloy groups, by major alloying elements,	
Copper	2***
Manganese	3***
Silicon	4***
Magnesium	5***
Magnesium and silicon	6***
Zinc	7***
Other elements	8***
Unused series	9***

In the 1*** group, the second digit indicates the modification in impurity limits or alloying elements. The last two of the four digits are the same as the last two digits in 99.** per cent and indicate the minimum aluminium percentage.

In the 2*** to 8*** groups, the second digit indicates the alloy modifications. The last two digits identify the different aluminium alloys in the group.

A letter after the four digits indicates a national variation.

Aluminium alloys were introduced to increase the strength of aluminium. The main alloying elements are copper, magnesium, manganese, silicon and zinc. Other elements, such as nickel and chromium, are also used. Chromium increases resistance to stress corrosion. Copper increases strength and reduces corrosion resistance, ductility and weldability. Magnesium reduces the melting point but increases corrosion resistance to sea water. Nickel increases strength in high temperature conditions. Silicon increases ductility and strength. Zinc increases strength.

1.2.2 Temper or condition (A.3)

Following the four-digit designation and a hyphen, the temper or condition of the aluminium or aluminium alloy is shown by a letter and possibly one or more numbers.

For the non-heat-treatable alloys, strength is increased only by strain-hardening but is decreased by heating. The temper is identified by the suffix letters as follows:

F As fabricated.
O Annealed.
H Strain-hardened.

The H is followed by two or more digits to show the final degree of strain-hardening. The first digit following the H has the following meaning:

H1 Strain-hardened only.
H2 Strain-hardened and partially annealed.
H3 Strain-hardened and stabilized.

A second digit following the H1 or H2 shows the degree of strain-hardening in ascending order of temper.

For heat-treatable alloys, enhanced strength is derived by one or two stages of heat treatment, namely solution heat treatment and precipitation heat treatment (ageing). With some alloys ageing occurs naturally, and the second formal heat treatment may not be needed. The heat-treatable condition is identified by suffix letters and numbers as follows (A.3.3):

O Annealed.
F As fabricated with no formal heat treatment.
T4 Solution heat-treated and naturally aged.
T5 Cooled from an elevated temperature shaping process and then artificially aged.
T6 Solution heat-treated and then artificially aged.
T8 Solution heat-treated, cold-worked and then artificially aged.

Some alloys still use the existing BS temper designation system.

Table A1 of BS 8118: Pt 1 lists the nearest foreign equivalents of 14 of the British Standards wrought and cast alloys referred to in Tables 2.1 and 2.2 of BS 8118: Pt 1. Table 2.1 lists five alloys and 14 conditions. Table 2.2 lists nine alloys and 26 conditions.

1.3 Permitted materials (2.2)

1.3.1 Standard materials (2.2.1.1)

Part 1 of BS 8118 covers the design of structures fabricated from the heat-treatable alloys of Table 2.1 and from the non-heat-treatable alloys of Table 2.2, commonly supplied to the specifications of BS 8118: Pt 2. Castings are not designed to BS 8118 unless there has been close consultation between the engineer and the casting manufacturer.

1.3.2 Heat-treatable alloys (2.2.1.1.2)

6082 is the commonest heat-treatable alloy because of its combination of good corrosion resistance and good physical properties. It is of medium strength, has a durability rating B, is usually used in the fully heat-treated condition and is used in both non-welded and welded structures. There is a loss of strength in the HAZ. 6082 is available in most forms: hollow and solid, extrusions, forgings, plates, sheets and tubes.

6061 is an alternative alloy to 6082 and has slightly improved formability and surface finish. It is of durability rating B, is available in extruded tube form and is mainly used for structures.

6063 combines moderate strength with a durability rating of B and good surface finish. There is a loss of strength in the HAZ. 6063 is available as extrusions, forgings and tubes and is especially used for architectural applications.

7020 is readily weldable, of durability rating C and is sensitive to environmental conditions. The performance of 7020 depends upon the control of its manufacture, including its composition and tensile properties, and of its fabrication. It is usually available only in rolled forms, simple extruded solids and hollow sections.

LM25 is a casting alloy, of durability rating B, with good corrosion resistance and mechanical properties.

1.3.3 Non-heat-treatable alloys (2.2.1.1.3)

Only strain-hardening enhances the strength of the non-heat-treatable alloys. These alloys are usually produced in plate and sheet form.

1200 is of durability rating A, with high ductility and very good corrosion resistance. It is available only in sheet and is used for architectural work.

3103 is of durability rating A, is similar to 1200, but is stronger and harder. It is available in sheet form and is used for building sheet and vehicle panelling.

3105 is of durability rating A and is used in the profiled building sheet market. It is available only in sheet form.

5083 is of durability rating A and is used for welded structures, plating and tank work. It welds readily without significant loss of strength and has high ductility. Long exposure to temperatures above 65°C in some adverse environments is to be avoided. It is available in plate, sheet, single extruded sections, drawn tube and forgings.

5251, 5154A and 5454 are all of durability rating A; they have good weldability and very good resistance to corrosive attack. They are available in plate, sheet and simple extrusions. Welded tubes of durability rating B are produced from 5251 and are used for handrails, ladders, etc.

LM5 is a medium-strength casting alloy of durability rating A having excellent finishing properties.

LM6 is of durability rating B and has high ductility, high impact strength and good corrosion resistance.

1.3.4 Non-standard tempers and conditions (2.2.1.2)

Non-standard tempers and conditions, plus other thicknesses of the alloys listed in Tables 2.1 and 2.2, are available, but consideration must be given to in-service behaviour, durability, resistance to crack propagation and to weldability. Table 2.3 gives the bolt and rivet materials together with their durability ratings. Table 2.4 gives the durability ratings for the filler metals used for MIG (metal inert-gas) and TIG (tungsten inert-gas) welding (2.2.3).

1.4 Mechanical, physical and strength properties (2.3)

Table 4.1 gives the limiting stresses for the heat-treatable alloys 6061, 6063, 6082 and 7020. Table 4.2 gives the limiting stresses for the non-heat-treatable alloys 1200, 3103, 3105, 5083, 5154A, 5251 and 5454. Tables 2.1 and 2.2 give the available forms, mechanical properties and temper conditions of the standard alloys (2.3.1).

As the mechanical properties of the alloys change with temperature, the design range of any aluminium alloy structure designed to BS 8118 is −50°C to +70°C, except for alloy 5083 where there will be a strength loss at temperatures over 65°C. Aluminium alloys melt within the range 550°C to 650°C (2.3.1).

The physical properties of the standard alloys are usually taken as constant and having the following values (2.3.2):

Coefficient of thermal expansion	23×10^{-6} per °C
Density	2710 kg/m^3
Modulus of elasticity	70,000 N/mm^2
Modulus of rigidity	26,600 N/mm^2

1.5 Durability and corrosion protection (2.4)

Normally the standard materials listed in Tables 2.1 to 2.4 do not need surface protection as their good corrosion resistance is due to the self-sealing film which forms on the surface of the metal. In mild environments, an aluminium surface retains its original appearance. In moderate industrial environments, the surface will darken and roughen. As the atmosphere becomes more aggressive, the discolouration and roughening increases with visible white powdery surface oxides. Added protection is then necessary, as it also is in coastal and marine environments. Alloys immersed in water also require special precautions. When surface attack does occur, corrosion/time curves usually follow an exponential form. There is rapid initial loss of reflectivity over six to 36 months, thereafter slowing down with little change over the next ten to 80 years. Tropical environments are, in general, no more harmful than temperate environments (2.4.1).

1.5.1 Alloy durability (2.4.2)

The alloys listed in Tables 2.1 to 2.4 are categorized into a descending order of durability, A, B, and C. The durability is used to determine the required protection. Where more than one alloy, including filler metals, is used, the protection should be in accordance with the lowest of their durability ratings.

1.5.2 Corrosion protection (2.4.3)

The requirement to provide corrosion protection when exposed to different environments is given in Tables 2.6 and 2.7 and described in Section 2.4.3 and Section 4 of BS 8118: Pt 2. When selecting the appropriate column of Table 2.6 for an atmospheric environment, it is advisable to study the precise conditions prevailing at the actual site. For hollow sections, the need to protect the internal surfaces may not be necessary if the internal void is sealed effectively.

Aluminium may require one of a number of protection systems as follows (Pt 2, 4.1):

1) Aluminium may be painted with either a non-bituminous system or a bituminous system (Pt 2, 4.2). Pre-coated aluminium and factory-applied finishes may be used, but with special care being taken over material heating. Additional protection at the joints may be necessary.
2) At metal-to-metal contact surfaces, such as bolted, bonded, pinned, riveted and welded joints, one of six protection procedures, numbered 0 to 5, may be specified (Pt 2, 4.3). Procedure 0 specifies no additional treatment, but 1 through to 5 specifies increasing protection. The requirements for additional protection are given in Tables 2.6 and 2.7 (2.4.3.2).
3) When aluminium is in contact with normal building materials such as concrete, masonry or plaster, the contacting surface should be coated with bituminous paint (2.4.3.3). If the alloy is embedded in concrete or in contact with soils, at least two coats of bituminous paint may be necessary. Additional consideration should be given to submerged contact, contact where water is constantly being replenished and where sea-dredged aggregates are used (2.4.3.3.2).
4) Timber is usually protected by preservatives or is primed and painted (2.4.3.3.3). When used in damp conditions, certain timber preservatives can affect the alloy and a sealant may be required. Further, some timbers, unless well seasoned, can also affect aluminium.
5) Aluminium of durability rating A should be used when immersed in contaminated water, fresh water or sea water. Tables 2.6 and 2.7 list the protections needed for fresh and sea water. Analysis of the water may also be necessary (2.4.3.3.5).
6) Some chemicals normally used in the building industry, such as cleaning materials, fungicides and mould repellents, when used in damp or wet conditions may affect aluminium. The aluminium may need to be protected. It is also necessary for building industry insulating materials to be checked for compatibility with aluminium where there are in-service damp and/or saline conditions (2.4.3.3.6).

1.6 Workmanship (Pt 2, 3)

The fabrication and assembly operations associated with structural aluminium are similar to those required for steel, but are affected by the alloy's lower mass, greater flexibility and larger dimensional changes because of temperature. These dimensional changes can affect marking out and assembly. It is necessary to ensure stability of the structure under all erection conditions (Pt 2, 3.1 and 3.4).

Aluminium should be stored in a dry place, off the ground, not in contact with other metals and away from other materials such as cement and timber (Pt 2, 3.2).

When cutting aluminium, whether by arc-cutting, machining or shearing, the tools must not be contaminated by other metals. Arc-cutting must first be checked by test. For the 7*** series alloys, residual stresses due to sheared edges must be removed by machining (Pt 2, 3.5).

Holes are made by drilling, drilling and reaming, or by punching. Punching is not used for the 2*** and 7*** series alloys (Pt 2, 3.6).

The manufacturer's recommendations should be followed where bending or forming is required (Pt 2, 3.7).

Washers must be used under all bolt heads and nuts. Steel bolts require galvanized steel washers, but for corrosion-resisting steel bolts the washers may be manufactured from the bolt material, pure aluminium or the same material as the member (Pt 2, 3.8.1).

Care must be taken with hot-driven steel rivets to avoid aluminium overheating (Pt 2, 3.8.3).

1.7 Welding (Pt 2, 3.9)

Welding should only be carried out under shop or simulated shop conditions. On-site welding should be avoided. The effect of distortion, warping and weld shrinkage must be considered, with strength members being welded using either the TIG or the MIG processes. The surfaces to be welded must be cleaned and welded within six hours.

The welding procedure for each type of joint shall be approved by the designer. The welders must be approved for each type of joint to BS 4871: Pt 2 and all production welds must be inspected and approved.

The three quality levels for each weld depends on the stressing requirements. The quality levels are as follows (Pt 2, 3.9.9.3):

1. minimum quality, when required, must be specified on the drawings,
2. normal quality, is assumed if the drawings specify no other quality and
3. fatigue quality, which is specified on the drawings by means of a 'Fat' class number and an arrow indicating the direction of stress fluctuation.

As strength loss may occur near a weld, the choice of filler and parent materials should be considered. Table 2.8 of Part 1 selects the correct filler wire for welded construction (2.5.3).

Unless fully tested, adhesives should not be used for the main structural joints. Adhesive can be used in secondary stressed conditions (Pt 2, 3.10).

1.8 Materials selection and availability (2.6, 2.7)

When choosing an alloy, it is necessary to consider availability, durability, formability, physical properties, strength and weldability (2.6).

Not all alloys given in Tables 2.1 and 2.2 are available in all the following product forms: castings, drawn tube, forgings, longitudinally welded tube, plate, sheet and strip plus the extruded sections, namely, complex and special sections, hollow sections, round and rectangular tube, plus solid bar and simple shapes. Table 2.9 indicates the combinations of alloys and product forms. Structural sections are usually produced by extrusion.

Many aluminium product forms and sizes are available from stock or are standard products manufactured to order. Others, including a very wide range of new sections which are produced by extrusion, are available by special arrangement. Minimum quantities may be applied and the additional charges may be small (2.7.2).

1.9 Design example 1

A beam consisting of a web plate welded to two flange plates is to be constructed using an aluminium alloy and condition that does not lose strength when welded. Choose a suitable sheet or plate alloy and condition.

Tables 2.1 and 2.2 give the approximate percentage reduction in alloy strength for each temper. The appropriate sheet or plate alloy and conditions, with no loss of strength due to welding, are as follows:

Table 2.1 Heat-treatable alloys:
6082-T4 and 7020-T4.
Table 2.2 Non-heat-treatable alloys:
5083-O, 5083-F, 5154A-O and 5454-O.

The HAZ softening factor k_z could also be used (4.4.2.1). The value of k_z is given values in Table 4.5. In certain circumstances, Appendix F should also be considered. For no strength reduction $k_z = 1.0$ (4.4.1). The appropriate plate alloys and condition combinations for $k_z = 1$ from Table 4.5 are as follows:

Heat-treatable alloys:
6082-T4 and in certain stress conditions 7020-T4 (4.4.2.2).
Non-heat-treatable alloys:
5083-O, 5083-F, 5154A-O, 5154A-F, 5254A-F, 5454-O and 5454-F.

Table 4.5 must be read in conjunction with Tables 4.1 and 4.2 as Table 4.5 does not fully differentiate the alloy condition and the product form. Table 4.1 shows that both 6082-T4 and 7020-T4 are available in sheet and plate. Table 4.2 shows that only 5083-O, 5083-F. 5154A-O and 5454-O are available in sheet and plate.

Care must also be taken as the approximate loss in strength due to welding given in Tables 2.1 and 2.2 does not always correspond to the appropriate k_z factors given in Table 4.5.

2 Design principles

2.1 Limit state design (3.1)

The limit states that need to be considered when designing an aluminium structure are:

1. durability (3.5),
2. fatigue (3.6),
3. serviceability limit state of deformation (3.4),
4. ultimate limit state of static strength (3.3) and
5. vibration (3.7).

The design will follow the guidance on calculations given in Sections 4, 5, 6 and 7 plus Appendices B to L inclusive of BS 8118: Pt 1. Testing in accordance with Section 8 can also be used to verify a proposed design.

2.2 Loading (3.2)

The structure should be designed to resist the loads as follows:

1. Dead loads: Structure self-weight and any permanently attached items.
2. Imposed loads: All statically or dynamically applied loads other than dead or wind loads.

3. Wind loads: Dynamic loads due to wind gusts. Wind loading can be the most severe load applied to the structure.
4. Temperature effects: Forces developed due to temperature changes.

The loads should be considered separately or in realistic combinations that will produce the most critical loadings. Explosions, fire, fluctuating loads, loads during assembly, support settlement, seismic forces and vehicular impact also need to be considered.

2.2.1 Nominal loading (3.2.2)

Nominal loads are used to check deformation, fatigue and vibration limit states. Nominal loads are determined from the relevant British Standards or are agreed between the designer and the client if no British Standards are relevant. The effect of internal building temperature and soil saturation loading must also be considered.

2.2.2 Factored loading (3.2.3)

Factored loads are the nominal loads multiplied by the overall load factor γ_f and are used to check the limit state of static strength. $\gamma_f = \gamma_{f1} \times \gamma_{f2}$ and makes allowance for accidental overload, variability in loading, etc. γ_{f1} is a partial load factor related to the type of load and has values between 0.8 and 1.33 (Table 3.1). γ_{f2} is a second partial load factor, related to load combinations and has ranges between 0.4 and 1.0 (Table 3.2). Alternative values of γ_{f1} and γ_{f2} may be used by agreement between the designer and the client, but care must be taken as other British Standards may define γ_{f1} differently.

2.2.3 Dynamic effects (3.2.4)

Dynamic effects are treated as imposed loads for γ_{f1}. Nominal loading due to dynamic conditions should be determined from an appropriate British Standard. If a British Standard is not appropriate, then a 'dynamic magnification factor' should be used and consideration given to the response of the structure. The response of high flexibility structures and structures capable of resonating has to be carefully considered. Additional calculations, structural damping and prototype testing may then be needed.

2.3 Static strength (3.3)

If the action effect under factored loading is less than the factored resistance, then the static strength criterion is satisfied. The action effect is the axial force, bending moment or shear force due to the application of the factored loads.

2.3.1 Factored resistance (3.3.3)

The factored resistance is the calculated resistance divided by the material factor γ_m. Sections 4, 5 and 6 give rules for establishing the calculated resistance, while Appendix B uses probability and statistics to establish the calculated resistance. γ_m takes into account the differences between the strengths of the test specimens and the strength of the actual structure. The value for γ_m is found in Table 3.3 and varies between 1.2 and 3. Different values of γ_m may be used by agreement between the client and the designer.

2.4 Deformation (3.4)

Structural deformation is acceptable if the elastic deflection due to nominal loading is less than the limiting deflection. The nominal loading may, in some circumstances of combined imposed loading, be reduced by multiplying by γ_{f2} taken from Table 3.2. Generally, the gross cross-section properties should be used for elastic deformation calculations, but for local buckling of slender sections a reduced section may have to be used (3.4.1).

The relevant British Standard, or an agreement between the client and the designer, provides the basis for determining the limiting deflection. Aluminium is three times more flexible than steel and, although BS 8118 recommends deflection limits of between L/360 to L/100 (Table 3.4), other larger deflection limits may be appropriate (e.g. for curtain wall mullions). L is the length between supports.

It is assumed that where components have been designed using Section 4 (Static design of members), there will be no significant permanent deformation due to nominal loading, but changes in dimensions of repeatedly assembled structures must be considered (3.4.2, 3.4.3).

2.5 Durability (3.5)

Tables 2.1 to 2.4 give the durability rating of the alloy groups. For the purposes of durability, a structure is considered satisfactory if the degree of exposure and the design life have been considered and a durable alloy used, which is protected in accordance with BS 8118: Pt 2.

2.6 Fatigue (3.6)

Where there are significant variations in load, the two fatigue limit states of:

a) stable crack growth and
b) total collapse under the (unfactored) design load spectrum,

should be considered (3.6.1). The predicted life, calculated in accordance with Section 7 (Fatigue), must not be less than the design life (3.6.2).

A fatigue life factor $\gamma_L > 1$ may be used to increase the (nominal) design life. The value of γ_L is influenced by:

a) possible changes in structure use,
b) the accuracy of the assumed loading spectrum,
c) the possible increase in crack growth and
d) whether lifetime loading records will be kept.

For example:

a) $\gamma_L = 1$ would give the maximum economy in design,
b) $\gamma_L = 1.5$ would reduce possible problems towards the end of the structure's fatigue life and
c) $\gamma_L = 4$ or 5 would be used if the future use of the structure is unknown. (Unknown future structure use is not a good basis for a design!)

The design stress range may also be divided by a fatigue material factor $\gamma_{mf} > 1$. The use of γ_{mf} is influenced by: i) the detail being in a very hostile environment or an environment where inspection is difficult, ii) the result of the failure of the detail and iii) whether there exist alternative load paths. A typical value for i) could be $\gamma_{mf} = 1.25$.

2.6.1 Stable crack growth (3.6.3)

Crack growth is assumed to be stable when it does not suddenly increase between inspections. The allowable limiting crack lengths, allowable rates of crack growth, the method of inspection and time between inspections should be agreed between the client and the designer.

2.7 Vibration (3.7)

Consideration should be given to the occurrence of undesirable vibrations under normal service conditions. Nominal loads are used. Fatigue failure may need to be considered.

2.8 Testing (3.8)

Components not designed to BS 8118: Pt 1 can only be accepted after successful testing in accordance with Section 8. The designer must be aware of the quality and limitations of the testing procedures and techniques. For example:

a) Have enough tests been carried out to allow a statistical analysis?
b) Will a component with high stress raisers require fatigue tests?
c) Is the structure to give 'damage-tolerant' or 'safe-life' performance?
d) Can the welding procedures on the test specimens be repeated during production?

2.9 Design example 1

An extruded symmetric I-section beam, 160 mm × 80 mm, has a span of 5 000 mm. The factored uniformly distributed dead load is 4.20 kN/m. The factored live load is 4.66 kN and can be placed anywhere on the span. As aluminium is three times more flexible than steel, the designer and the client have agreed a limiting deflection of L/100 = 5 000 mm/100 = 50 mm (3.4.1). The beam will be used in a rural atmospheric environment. Determine if the beam has a deflection less than the limiting deflection, also determine the required minimum durability rating.

From Table 5 of BS 1161 for a 160 mm x 80 mm I-section, $A = 2\,820\ mm^2$ and $I_x = 1\,170\ mm^4 \times 10^4$. Table 2.5 gives $E = 70\,000\ N/mm^2$. Table 2.6 gives C as the durability rating.

The loads are considered either separately or in realistic combinations (3.2.1).

The factored loads = γ_f times the nominal loads, where $\gamma_f = (\gamma_{f1} \times \gamma_{f2})$.

γ_{f1} = 1.2 for the dead load and
= 1.33 for the imposed load (Table 3.1).

γ_{f2} = 1.0 for the dead load and
= 1.0 for the imposed load (Table 3.2).

The nominal loads are required for deformation (3.2.2).

The nominal dead load = 4.2 kN/m/(1.2 × 1.0) = 3.5 kN/m.
The nominal live load = 4.66 kN/(1.33 × 1.0) = 3.5 kN.

The agreed limiting deflection is 50 mm. The actual deflection is calculated using the nominal loads (3.2.2) and the gross section (3.4.1).

For the UDL, the maximum deflection

$= (5wl^4)/(384EI)$
$= (5 \times 3.5\ \text{N/mm} \times (5\,000\ \text{mm})^4)/(384 \times 70\,000\ \text{N/mm}^2 \times 1\,170\ \text{mm}^4 \times 10^4)$
$= 34.78$ mm.

For the single load, the maximum deflection

$= (Wl^3)/(48EI)$
$= (3\,500\ \text{N} \times (5\,000\ \text{mm})^3)/(48 \times 70\,000\ \text{N/mm}^2 \times 1\,170\ \text{mm}^4 \times 10^4)$
$= 11.13$ mm.

Making a total deflection of 45.91 mm (34.78 mm + 11.13 mm), which is less than the allowable deflection of 50 mm, therefore satisfactory. If the total deflection had been greater than the allowable deflection and there had been a combination of imposed loads, then the combined imposed loading could legitimately be reduced by multiplying the nominal loads by γ_{f2} and not by γ_f (3.4.1).

2.10 Design example 2

A structural component (a roof beam) can be subjected to the following nominal loads (there are no temperature effects (3.2.1)):

a) a dead load 0.95 kN/m^2,
b) a snow loading 0.75 kN/m^2 and
c) a wind load of plus or minus 1.20 kN/m^2.

Determine the worst combination, per square metre, of the factored loads.

Nominal loads are used for the limit states of deformation, fatigue and vibration (3.2.2). Factored loads, which are equal to γ_f times the nominal loads, are used to check the limit state of static strength, $\gamma_f = \gamma_{f1} \times \gamma_{f2}$ (3.2.3).

Using Table 3.1,

$\gamma_{f1} = 1.2$ for direct dead load,
$\gamma_{f1} = 0.8$ for dead load if countering uplift or overturning,
$\gamma_{f1} = 1.33$ for imposed load and
$\gamma_{f1} = 1.2$ for wind load.

Using Table 3.2,

$\gamma_{f2} = 1.0$ for dead load,
$\gamma_{f2} = 1.0$ for the most severe imposed load or wind load (in this case the wind load) and
$\gamma_{f2} = 0.8$ for the second most severe imposed load or wind load (in this case the snow load).

The worst factored load combination is:

$$\begin{aligned}&\text{dead} + \text{wind} + \text{snow}\\ &= (0.95\ \text{kN/m}^2 \times 1.2 \times 1.0) + (1.20\ \text{kN/m}^2 \times 1.2 \times 1.0) + (0.75\ \text{kN/m}^2 \times 1.33 \times 0.8)\\ &= 3.378\ \text{kN/m}^2.\end{aligned}$$

For uplift, the worst factored load combination is:

$$\begin{aligned}&\text{dead} + \text{wind}\\ &= (0.95\ \text{kN/m}^2 \times 0.8 \times 1.0) - (1.20\ \text{kN/m}^2 \times 1.2 \times 1.0)\\ &= -0.68\ \text{kN/m}^2\ \text{(uplift)}.\end{aligned}$$

As the roof beam can be subjected to uplift, it must be securely held down. The reversal in loading due to the wind would need to be considered further if the beam had high flexibility and a natural period of vibration similar to that of the imposed load (3.2.4).

If deformation is to be considered, the overall load factor of γ_f for the combinations of imposed loads may be replaced by γ_{f2} (3.4.1). This is to avoid unduly small limiting deflections.

3 Static design of members (4)

3.1 Introduction

The limit states of deformation (4.9) and static strength (4.1.2) must be satisfied for all members formed from, or formed from combinations of, extrusions, plates, tubes and sheets. Castings are not included. Where design curves are used, the formulae used to obtain the design curves, which are given in Appendix K of BS 8118: Pt 1, may be used instead (4.1.1). The use of the appendices may lead to lighter designs (4.1.4).

3.2 Limit state of static strength (4.1.2)

The action effect under factored loading must not be greater than the factored resistance. The rules to be used to determine the resistance to different actions are found in Sections 4.5, 4.6 and 4.7. For beams, the resistance to moment and shear is described in Section 4.5; for struts, the resistance to axial compression is given in Section 4.7; and for ties, the resistance to axial tension is found in Section 4.6. The combination of axial and bending loads is given in Section 4.8 (4.1.2).

The resistance calculations assume the limiting stresses p_o, p_a, and p_v as given in Table 4.1 for heat-treatable alloys and in Table 4.2 for the non-heat-treatable alloys (4.2). Appendix D covers materials not in these two tables. The limiting stresses are modified by the material factor γ_m as given in Table 3.3 (4.1.2). The determination of p_s is dealt with later (4.5.6.5 and 4.7.6).

The resistance to local buckling of certain slender members under axial compression or moment compression may be reduced due to their cross-section slenderness. Some sections are classified in terms of their susceptibility to local buckling (4.1.2).

3.3 Heat-affected zones (HAZs) (4.1.3)

Even small welded joints contain very high residual stresses and may very much weaken the resistance of a member in the heat-affected zone (HAZ). Welds should, if possible, be located in areas of low stress. The effect, extent and severity of the softening due to a HAZ, that must be considered, are given in the rules in Section 4.4. These design considerations do not apply to parent metal in the conditions O or T4, nor to the F condition when the design is based on the O condition properties.

3.4 Section classification and local buckling (4.3)

The resistance of some members is classified in terms of their susceptibility to local buckling and the slenderness of the individual elements of the member need to be considered (4.3.1.1).

The rules identify three basic types of thin-walled elements:

a) curved internal elements,
b) flat internal elements and
c) flat outstand elements.

These three element types are not usually longitudinally stiffened, i.e. they are unreinforced, as shown in Figure 4.1a. The inclusion of longitudinal stiffening ribs or lips, i.e. they are reinforced, as shown in Figure 4.1b, will greatly increase the stability of the flat elements (4.3.1.2). The buckling of shear webs is considered later in Section 4.5.3.3 or in Section 5 (4.3.1.3).

3.4.1 Slenderness parameter β *(4.3.2)*

The susceptibility to local buckling in a beam or in a strut depends upon β which is defined in several ways (4.3.2.1). For unreinforced flat elements β depends upon b/t and d/t (4.3.2.2). b and d are the flat element width and, where relevant, are measured to the springing of a fillet or to the toe of a weld (4.3.2.2), where:

- b is the width of the element generally,
- d is the depth of a web element in a beam,
- g is the stress gradient coefficient found from Figure 4.2,
- t is the element thickness,
- y_c is the distance from the neutral axis of the gross section to the more heavily compressed edge taken as positive towards the compression side and
- y_o is the distance from the neutral axis of the gross section to the less heavily compressed edge or to the edge in tension taken as positive towards the compression side.

a) For an element under uniform compression: $\beta = b/t$.
b) For an element under a stress gradient:
   1) An internal element with a stress gradient that gives a neutral axis at the centre: $\beta = 0.35b/t$ or $\beta = 0.35d/t$
   2) For any other stress gradient: $\beta = gb/t$ or $\beta = gd/t$.

y_c and y_o are normally measured from the elastic neutral axis. When checking whether a section is fully compact, measuring from the plastic neutral axis is permissible (4.3.2.2).

For reinforced flat elements, two buckling modes need to be considered as shown in Figure 4.3. They are as follows (4.3.2.3):

a) mode 1: the reinforced element buckles as a unit taking the reinforcement with it and
b) mode 2: the sub-elements making up the reinforced elements themselves, buckle as individual elements, the junctions between them staying straight.

For mode 1 buckling under uniform compression, β is found as follows:

a) If the reinforcement consists of a single-sided rib or lip of the same thickness t as the element, located as shown in Figure 4.4, it is considered as standard reinforcement and:

$\beta = hb/t$

The value of h is found from Figure 4.4, with the measurement of c, the clear depth of the rib or lip measured to the surface of the plate, being defined as in Figure 4.4.

b) With any other shape of reinforcement β is found by replacing the actual reinforcement with an equivalent lip or rib of the standard form and then following a) above.
The value of c to be used must ensure that the second moment of area about the mid-plane of the plate is the same for both the equivalent rib and for the true reinforcement.

c) For any cases not covered by a) and b) above:

$$\beta = (b/t)(\sigma_{cro}/\sigma_{cr})^{0.4}$$

where, assuming simple edge supports:

σ_{cr} = the elastic critical stress of the element with reinforcement and
σ_{cro} = the elastic critical stress of the element without reinforcement.

For mode 1 buckling under a stress gradient:

$$\beta = (b/t)(\sigma_{cro}/\sigma_{cr})^{0.4}$$

but where σ_{cro} and σ_{cr} now relate to the stress at the more heavily compressed edge of the element (4.3.2.3).

For a reinforced outstand element in a beam, subjected to mode 1 buckling, with the reinforcement in the form of an outwardly facing lip, forming all or part of the compression flange, the reinforcement is ignored in classifying the section (4.5.2.1).

The mode 2 buckling parameter β is found for each sub-element as if the sub-elements were unreinforced flat elements (4.3.2.3).

3.4.2 Curved internal elements (4.3.2.4)

Provided $R/b \geq 0.1b/t$, then for a shallow curved element under uniform compression:

$$\beta = \frac{b}{t\,(1 + [0.006b^4/\{R^2t^2\}])^{0.5}} \text{ where:}$$

b = the developed width of the element at mid-metal,
R = the radius of curvature to the mid-metal and

t = the thickness.

Provided $R/b \geq 0.1b/t$, for a shallow curved element under a stress gradient:

$$\beta = \frac{gb}{t(1 + [0.006b^4/\{R^2t^2\}])^{0.5}} \text{ where:}$$

g = the stress gradient coefficient found from Figure 4.2 (4.3.2.4).

3.4.3 Round tubes (4.3.2.5)

For round tubes, whether under axial compression and/or bending,

$\beta = 3(D/t)^{0.5}$ where:

D = the diameter to the mid-metal and
t = the thickness.

3.5 Section classification (4.3.3)

Except for elements stressed wholly in tension, the individual elements making up the section are classified, with the least favourable element determining the classification of the section (4.3.3.1).

3.5.1 Beam and strut sections (4.3.3.2 and 4.3.3.3)

The classification for the section of a beam resisting a moment or a strut resisting an axial force is as follows (4.3.3.2):

a) Moment resistance:
   1) Fully compact, then local buckling can be ignored.
   2) Semi-compact, then the section can develop a moment equal to p_o times the elastic section modulus.
   3) Slender, then premature local buckling at an extreme fibre occurs below a stress of p_o and thus reduces the moment resistance.

b) Axial compression resistance:
   1) Compact, then local buckling can be ignored.
   2) Slender, where local buckling lowers the resistance.

p_o is the limiting stress for bending and for overall yielding as given in Tables 4.1 and 4.2 (4.3.3.2).

3.5.2 Combined actions on sections (4.3.3.3)

For sections required to carry biaxial bending or axial force with simultaneous bending, the section is given a single classification of fully compact, semi-compact or slender, in accordance with Sections 4.3.3.2, 4.3.3.4a or 4.8.2.1. When obtaining the axial resistance, if the section is either fully compact or semi-compact, then no reduction has to be made for local buckling (4.8.2.1).

For each element, the value of g required to modify β is found from Figure 4.2 and must correspond to the stress pattern that is produced when all the moments and the axial force are applied simultaneously. Figure 4.2 requires the quantities y_o and y_c. These quantities are usually found by using the elastic neutral axis of the gross section subjected to the combined actions. The plastic neutral axis may be used if a check is being made on whether a section is fully compact. Normally y_o and y_c will have different signs, but if the elastic neutral axis lies outside the section, they will have the same sign (4.8.2.1).

If the flange elements are under-stressed, the elements may be more favourably classified using Section 4.3.3.5, provided y_c and y_o relate to the stress pattern under the combined loading (4.8.2.1).

3.5.3 Element classification (4.3.3.4)

Individual elements are classified as follows (4.3.3.4):

a) Beam elements: moment resistance.
   1) $\beta \leq \beta_1$ fully compact
   2) $\beta_1 < \beta \leq \beta_o$ semi-compact
   3) $\beta > \beta_o$ slender
b) Strut elements: axial resistance.
   4) $\beta \leq \beta_o$ compact
   5) $\beta > \beta_o$ slender

where the values of β_o and β_1 are found from Table 3.1 overleaf and are related to internal and outstand elements, plus whether an element is unwelded or welded (4.3.3.4).

If an element has a weld at an edge or at any point in its width, it is considered as welded. The classification of welded elements is independent of the extent of the HAZ. However, if the stability of a member at a particular cross-section is being considered and there is no weld at that section, the member is considered as unwelded.

For under-stressed flange elements in a member, a more favourable classification may be used, if the actions are: a) bending or b) bending

Table 3.1 Limiting values of β

Elements	β_0 *Unwelded*	β_0 *Welded*	β_1 *Unwelded*	β_1 *Welded*
Outstand elements	7ϵ	6ϵ	6ϵ	5ϵ
Internal elements	22ϵ	18ϵ	18ϵ	15ϵ

$\epsilon = (250/p_o)^{0.5}$ where
p_o is the limiting stress found from Tables 4.1 or 4.2 (4.3.3.4).

Source: Reproduced with permission from BS 8118: Pt 1

with axial force, that are both parallel to the axis of bending, and less highly stressed than the most severely stressed fibres in the section (4.3.3.5). In this case, the value of ϵ in Table 4.3 becomes:

$\epsilon = ([250y_1]/[p_o y_2])^{0.5}$ where:

y_1 = the distance from the elastic neutral axis of the gross section to the most severely stressed fibres and
y_2 = the distance from the elastic neutral axis of the gross section to the element.

When checking to determine if a section is fully compact, it is permissible to use the plastic neutral axis rather than the elastic neutral axis.

3.6 Local buckling (4.3.4)

Local buckling in slender members is allowed for by substituting an effective section for the actual section. For a slender element partly or wholly in compression, the uniform thickness is reduced by a local buckling coefficient k_L. Non-uniform thickness is not considered (4.3.4.1).

The value of k_L is found for the different elements of the section, from the appropriate curve of Figure 4.5, selected using Table 3.2 opposite (4.3.4.2).

To select the correct curve, the value of β/ϵ is required. Generally, $\epsilon = (250/p_o)^{0.5}$, but $\epsilon = (250y_1)/(p_o y_2)^{0.5}$ can be used if the member being considered is a slender compression flange element of a beam, which lies nearer to the neutral axis than does the extreme fibre tensile

Table 3.2 Local buckling curve selection

Elements		*Unwelded*	*Unwelded*
Flat outstand elements		curve A	curve B
Flat or curved internal elements		curve C	curve D
Round tubes	The lower of:	curves C and E	curves D and E

Source: Reproduced with permission from BS 8118: Pt 1

material. The element must be substantially parallel to the axis of bending (4.5.2.3).

If an element has a weld at an edge or at any point in its width, it is considered as welded. However, if the stability of a member at a particular cross-section is being considered and there is no weld at that section, the member is considered as unwelded.

For reinforced flat elements, the more critical of buckling modes 1 or 2 must be used. For mode 1, k_L is applied to the basic plate thickness and to the area of reinforcement (4.3.4.2).

Where sections carry biaxial bending, or simultaneous axial force with bending, and have been classified as slender ($\beta > \beta_o$), each individual resistance must be found, using the relevant effective section for the action being considered. For example, k_L for each element under axial resistance relates to a value of β, corresponding to a uniform stress ($g = 1$). The value of β, for the moment resistance, relates to the stress patterns in the element, due to the section being under pure bending (4.3.4.3 and 4.8.2.2).

3.7 HAZ softening adjacent to welds (4.4)

Welding softens the area in the immediate vicinity of the weld. This softening can be reduced by the artificial ageing of the weld or by using a lower interpass temperature T_o (F.1.2). The maximum interpass temperature is 80°C for 7*** series alloys and 100°C for all other alloys. The softening affects 0.2 per cent proof stress to a greater extent than it affects the tensile strength. The worst effects of the welding are immediately around the weld but these effects rapidly attenuate. The effects of the weld are approximated within a zone, called the heat-affected zone (HAZ), by reducing the strength properties for the HAZ by k_z, the strength factor. The HAZ is defined as a distance z in any direction from the weld. Outside the HAZ, the full parent properties apply (4.4.1).

The value of k_z is found from Table 4.5. It is also found from

Appendix F entitled 'HAZs adjacent to welds'. Appendix F is usually used when resistance of a member is governed by p_a or p_v rather than p_o and may give more favourable values of k_z. k_z, in Appendix F, Tables F1 and F2, is redesignated as k'_z and applies for:

a) local failure in tension members (4.6.2.3),
b) local squashing in struts (4.7.7) and
c) shear in beams (4.5.3).

Table F2 gives values for the general determination of k_z and k'_z for materials not found in Tables 4.5 and F1 (F.2.2 and 4.4.2.1).

Table 4.5 gives two values of k_z for the 7020 material, namely A and B, which depend upon the nature of the stress acting on the HAZ material:

A is used when a tensile stress is acting transversely to the axis of a butt or fillet weld and
B which is used for all other stress conditions.

The value of 'A' can be increased, but is dependent upon the degree of thermal control exercised during fabrication, as discussed in Appendix F (4.4.2.2).

For welds in 7*** alloys, with the tensile stresses acting transversely to the weld axis, $k_z = k'_z$ and equals the A value. The B value is used for all other stress conditions (F2.3). More favourable values of 'A' can be used in certain circumstances (F2.3).

Table 4.5 assumes that for the values of k_z to apply to the heat-treatable alloys, the welded components will be held at a minimum of 15°C for at least three days for the 6*** series alloys and at a minimum of 15°C for at least 30 days for the 7*** series alloys (4.4.2.3). Neither series of alloys should be loaded within 24 hours of welding. If the 6*** or the 7*** series alloys are to be loaded before the three or 30 days respectively, then k_z is reduced by the reduction factor f as follows for:

6*** series alloys $f = 0.9 + 0.1([n - 1]/2)^{0.5}$ and for
7*** series alloys $f = 0.8 + 0.2([n - 1]/29)^{0.5}$ where

n is the time in days between welding and loading.

3.7.1 The extent of the HAZ (4.4.3)

The extent of the HAZ may be found experimentally as described in Appendix F, section F.5. Normally the HAZ is assumed to extend to a distance z in any direction from the weld as shown in Figures 4.6a to 4.6j, as follows:

a,	transversely from the centre-line of an in-line butt weld,
b, c, and d,	at butt welds used in tee, corner or cruciform joints, transversely from the point of intersection of the welded surfaces,
e, f, g, and h,	at fillet welds transversely from the point of intersection of the welded surfaces and
i, j,	in any radial direction from the end of a weld (4.4.3.1).

The HAZ boundaries are assumed to be straight lines normal to the metal surface, but for thick materials a curved boundary of radius z may be assumed as shown as B in Figure 4.6i. z is found from the following equation (4.4.3.2):

$$z = \alpha \eta z_o$$

where
z_o is the basic value as determined from Section 4.4.3.3,
α is the modifying factor for the extent of the HAZ to allow for elevated temperature (4.4.3.4) and
η is the modifying factor for the extent of the HAZ to allow for increased heat build-up (4.4.3.5).

A more favourable equation:

$$z = \delta \alpha \eta z_o$$

may be used, where normally
$\delta = 1$, but may be reduced as follows (F.3.2):
$\delta = 0.75$ for a joint that has three or more valid heat paths from it or
$\delta = (1.5 + [1.3L/z_o])/(3 + [L/z_o])$ for a straight joint of length L, where L is less than $5z_o$.

If both values of δ apply, the lower value is used (F.3.2).

Assuming complete interpass cooling of an isolated weld laid on unheated material, the basic value z_o (mm) is the lower of the two values given for z_o (4.4.3.3):

	7*** series alloys
In-line butt welds	$z_o = 30 + (t_A/2)$ or $4.5t_A$
All other types of butt weld and all types of fillet weld	$z_o = 30 + (t_A/2)$ or $4.5(t_B)^2/t_A$

	All other alloys
In-line butt welds	$z_o = 20 + (t_A/3)$ or $3t_A$
All other types of butt weld and all types of fillet weld	$z_o = 20 + (t_A/3)$ or $3(t_B)^2/t_A$

where:
t_A is the lower of $0.5(t_B + t_C)$ or $1.5t_B$,
t_B is the thickness of the thinnest element connected by the welding and
t_C is the thickness of the thickest element connected by the welding (4.4.3.3).

α is a modifying factor to allow for an elevated temperature in the material at the start of a weld pass (4.4.3.4). The values for α are given in Table 4.6 and vary between 1.0 and 2.0 for fabrication complying with BS 8118: Pt 2 (4.4.3.4). In certain cases lower values of α can be used if stricter thermal control than that required in BS 8118: Pt 2 is applied, or if $t_C \leq 25$ mm and the total area of all the weld deposits in the joint are known (F.3.3).

η is a modifying factor to allow for increased heat build-up due to the proximity of a free edge or edges, or other welding close by. η has one of the following values (4.4.3.5):

$\eta = 1$ where there are at least two valid heat paths from a joint,
$\eta = 1.33$ where there is only one valid heat path from a joint and $t_C > 25$ mm or
$\eta = 1.50$ where there is only one valid heat path from a joint and $t_C \leq 25$ mm.

A valid heat path has $h \geq h_1$ where:
h is the distance to a free edge, measured at mid-thickness and following the section profile, or half the distance to the adjacent weld (Figure 4.7 shows a typical heat path measurement),
$h = 4.5\alpha z_o$ for the 7*** series alloys and
$h = 3\alpha z_o$ for all other alloys.

If the thermal control is more strictly controlled than specified in BS 8118: Pt 2, then η may be read from Figure F.1 (F.3.4).

Where the HAZs from two joints overlap, the extent of the outer side HAZ is unaltered (4.4.3.6).

3.8 Design example 1

Determine the factored tension resistance P_{RS} of the butt-welded I-section shown in Figure 3.1 below. The alloy is 7020-T6 plate. The section

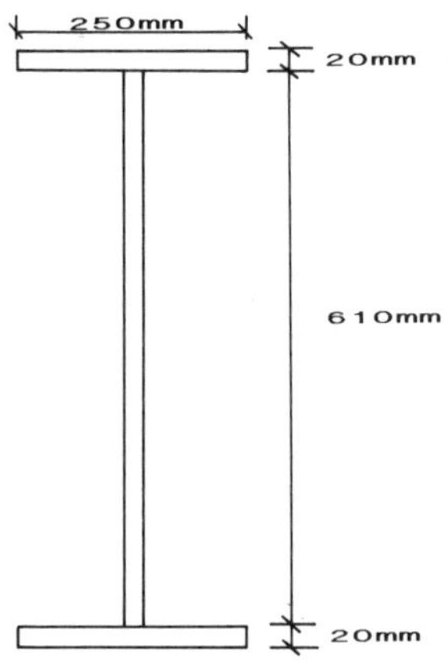

a) The I-section

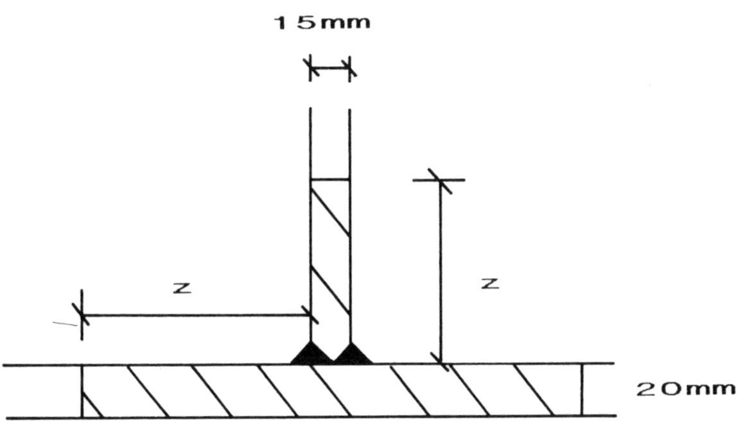

b) Extent of the HAZ

Figure 3.1 The I-section and the extent of the HAZ

is free from the effects of end connections and occasional holes. HAZ effects must be considered. Fabrication is to BS 8118: Pt 2 and the recovery time after welding as given in Section 4.4.2.3 is followed.

Table 4.1 of BS 8118: Pt 1 shows that alloy 7020-T6 is available in plate and sheet up to 25 mm in thickness, with $p_o = 270\ \text{N/mm}^2$ and $p_a = 295\ \text{N/mm}^2$. Table 3.3 gives $\gamma_m = 1.2$ for welded members. If 7020 had been used in the T4 condition, the HAZ effect could have been ignored (4.1.3), but p_o (160 N/mm^2) and p_a (205 N/mm^2) would have been reduced.

P_{RS} is the lesser of the values of either (4.6.2.1):

a) general yielding along the member (4.6.2.2b) or
b) local failure at a critical section (4.6.2.3b).

For general yielding, $P_{RS} = p_o A_e/\gamma_m = (270\ \text{N/mm}^2 \times A_e\ \text{mm}^2)/1.2$ as the flange and web have longitudinal butt welds.

A_e, the effective section area, is determined using Sections 4.4.2 and 4.4.3.

For 7020-T6 HAZ material under longitudinal stress, the B value in Table 4.5 is used as the butt weld has longitudinal stress (4.4.2.2b) and gives $k_z = 0.8$.

To determine z Figure 4.6 is used, where
$z = \alpha\eta z_o$ (4.4.3.2). z_o is the lower of
$z_o = 30 + (t_A/2)$ and
$z_o = (4.5t_B{}^2)/t_A$, where $t_B = 15$ mm and $t_C = 20$ mm. t_A is the lesser of
$0.5(15 + 20) = 17.5$ mm or
$1.5 \times 15 = 22.5$ mm, giving
$t_A = 17.5$ mm.
z_o is the lower of $30 + (17.5/2) = 38.75$ mm or $([4.5 \times 15^2]/20) = 50.63$ mm, giving
$z_o = 38.75$ mm.
$\alpha = 1.5$ from Table 4.6, case R, with $t_c \leq 25$ mm (4.4.3.4).

Regarding η, the number of heat paths must be considered. For a valid heat path, $h \geq h_l$ (4.4.3.5). For the 7*** series alloys,
$h_l = 4.5\alpha z_o$ giving
$h_l = 4.5 \times 1.5 \times 38.75$ mm
$= 261.56$mm.

Using Figure 4.6b and Figure 4.7, $h = (250\ \text{mm}/2) = 125$ mm for the flange and is thus not a valid heat path (4.4.3.5). For the web, we require half the distance to a nearby weld,
$h = (650\ \text{mm} - 20\ \text{mm} - 20\ \text{mm})/2$
$= 305$ mm, which is a valid heat path.

There is only one heat path where h (305 mm) $\geq h_1$ (261.56 mm) and with t_C (20 mm) $\leq$ 25 mm, gives
$\eta = 1.5$ (4.4.3.5) then
$z = 1.5 \times 1.5 \times 38.75$ mm
$= 87.19$ mm.

The area of the softened zone is:
$2 \times ([87.19 \text{ mm} \times 15] + 20 \text{ mm} \times [2 \times 87.19 \text{ mm} + 15 \text{ mm}]) = 10\,191 \text{ mm}^2$, with
$A = (2 \times 250 \text{ mm} \times 20 \text{ mm}) + 15 \text{ mm} \times (650 \text{ mm} - 20 \text{ mm} - 20 \text{ mm})$
$= 19\,150 \text{ mm}^2$ giving
$A_e = 19\,150 \text{ mm}^2 - ([1.0 - 0.8] \times 10\,192 \text{ mm}^2)$
$= 17\,112 \text{ mm}^2$.
Therefore $P_{RS} = (270 \text{ N/mm}^2 \times 17\,112 \text{ mm}^2)/1.2 = 3\,850.2$ kN.

For local failure:
$P_{RS} = p_a A_{ne}/\gamma_m$
$= (295 \text{ N/mm}^2 \times A_{ne} \text{ mm}^2)/1.2$. As there are no holes, $A_{ne} = A_n$ giving
$P_{RS} = (295 \text{ N/mm}^2 \times 17\,112 \text{ mm}^2)/1.2$
$= 4\,206.7$ kN.

Thus $P_{RS} = 3\,850.2$ kN.

4 Static design of beams (4.5)

4.1 Introduction

For beams and plate girders, it is usually necessary to check for the following (4.5.1.1):

1) lateral torsional buckling (4.5.6),
2) moment (4.5.2, 4.5.4 and Appendix E),
3) shear (4.5.3) and
4) web bearing (4.5.5).

The checking of 2) and 3) is more usual than the checking of 1) and 4).
For the moment check, $M \leq M_{RS}$, where:

M is the moment due to factored loading and
M_{RS} is the factored moment of resistance of the section and may need to be reduced if there is coincident shear.

For the shear check, $V \leq V_{RS}$, where:

V is the shear force due to factored loading and
V_{RS} is the factored shear force resistance.

Plate girders with slender stiffened webs can be designed as beams using Section 4.5, but they may be more efficient if designed as plate girders using Section 5.4 (4.5.1.2).

Beams subjected to a combination of loads, such as an axial load, plus a moment, or simultaneous bending about the principal axes, should be designed using Sections 4.8 (4.5.1.3 and 4.5.1.4).

4.2 Uniaxial moment resistance of the section (4.5.2)

The section of a beam resisting a moment is classified as one of the following using Section 4.3.3:

1) fully compact, where local buckling can be ignored,
2) semi-compact, where the section can develop a moment equal to p_o times the elastic section modulus or
3) slender, where premature local buckling at an extreme fibre occurs below a stress of p_o and thus reduces the moment resistance (4.3.3.2).

p_o is the limiting stress for bending and for overall yielding.

If a reinforced outstand element forms any part of the compression flange, then the reinforcement, if it is an outward-facing lip, is ignored in classifying the section (4.5.2.1).

4.3 Basic calculations (4.5.2.2)

M_{RS} is the factored moment of resistance and in the absence of shear is found as follows (4.5.2.2):

Unwelded,	fully compact	$M_{RS} = (p_oS_n)/(\gamma_m)$
	semi-compact	$M_{RS} = (p_oZ_n)/(\gamma_m)$
	slender, the smaller of	$M_{RS} = (p_oZ_e)/(\gamma_m)$ or $(p_oZ_n)/(\gamma_m)$
Welded,	fully compact	$M_{RS} = (p_oS_{ne})/(\gamma_m)$
	semi-compact	$M_{RS} = (p_oZ_{ne})/(\gamma_m)$
	slender, the smaller of	$M_{RS} = (p_oZ_e)/(\gamma_m)$ or $(p_oZ_{ne})/(\gamma_m)$

where:

S_n is the plastic modulus of the net section,

S_{ne} is the plastic modulus of the net effective section,

Z_e is the elastic modulus of the effective section,
Z_n is the elastic modulus of the net section,
Z_{ne} is the elastic modulus of the net effective section,
p_o is the limiting stress as defined in Tables 4.1 and 4.2 and
γ_m is the material factor as given in Table 3.3.

The effective section includes: a) the reduced thickness around the welds to allow for HAZ softening and local buckling but b) with no deduction for holes.

The net effective section includes: a) the deduction for holes and b) the reduced thickness around the welds to allow for HAZ softening.

The net section includes the deduction for the holes.

For semi-compact and slender sections, the moment resistance can be based on the elasto-plastic stress patterns (Appendix E), provided the more favourable assumed thickness of the section used for welded elements, as described in Section 4.5.2.3, is not applied (4.5.2.2).

4.3.1 Elasto-plastic stress pattern (E.1)

The elasto-plastic method has advantages for sections where a critical element for local buckling has a β value in the region of β_o and where the outermost tension fibres are further from the neutral axis than the compression fibres (Appendix E). For example, where an element of a critical compression flange is closer to the neutral axis than the furthest tension material, or where a critical web element, due to a tongue plate, stops some distance from the furthest compression face (Figure E.1).

An elasto-plastic stress pattern is constructed that assumes a definite material yield point at p_o. The calculations assume an effective net section, using a reduced thickness to cover HAZ softening. A deduction is made for holes. There is no thickness reduction for local buckling and the more favourable assumed thickness for welded elements is not applied (4.5.2.3). The element is classified as compact, semi-compact or slender (E.1, E.2.1 and 4.3.3.4), with $\epsilon = (250/p_o)^{0.5}$.

The limiting compressive stress p_m, for the slender element, is:

$$p_m = k_L p_o$$

where k_L is the local buckling factor (4.3.4.2), but with $\epsilon = (250/p_o)^{0.5}$.

For longitudinally stiffened webs, a series of values of p_m will be found due to the individual values of k_L of the various sub-panels. The k_L values are found by taking the correct width and stress pattern for each sub-panel to determine its β value (5.4.4.1). Further, the force in compression should balance the force in tension. Also, the extreme fibre tension stress must not exceed p_o, with the plasticity spreading in from that face. The maximum compression stress is the limiting value p, where:

$p = p_o$ for a fully compact or semi-compact element and
$p = p_m$ for a slender element.

If the compression flange is fully compact the plasticity is allowed to spread in from the compression face. Further, if the compression flange is fully compact and has a tongue plate next to the compression edge of a semi-compact or slender web, the distance the plasticity may spread in from the compression face is limited (E.2):

1) for a semi-compact web, the plasticity must not extend beyond the compression edge of the web and
2) for a slender web, $p = p_m$ for the web.

When a section is fabricated from components of differing strength (i.e. a hybrid section), the strain distribution will still be linear across the whole section. This implies:

1) either side of the neutral axis, the stress varies linearly in the elastic zone,
2) there will be a step-change in the stress, in the plastic zone, between materials with different values of p_o and
3) if the change of materials is at the same point as the junction between the elastic and the plastic zones, there will be a step-change in the stress (E.2.2).

The factored moment of resistance, M_{RS}, is as follows:

$$M_{RS} = M_u/\gamma_m$$

where M_u is the ultimate moment as determined using Sections E.3.2 or E.3.3 and γ_m is the material factor as given in Table 3.3.

For slender sections the stress patterns described earlier in this section are used to determine M_u (E.2.2).

For semi-compact sections:

$$M_{RS} = M_u/\gamma_m$$

where

$$M_u = M_{us} + ([\beta_o - \beta]/[\beta_o - \beta_1]) \times (M_{uf} - M_{us})$$

where:

M_{us} is the moment corresponding to the stress pattern adopted (E.2.1),
M_{uf} $= Sp_o$,
S is the plastic section modulus of the gross section, with no reduction for HAZ, holes or local buckling,
β is the β value of the critical element,
β_o is the semi-compact limiting value of β (Table 4.3) and
β_1 is the fully compact limiting value of β (Table 4.3).

In calculating M_{us}, if the compression flange is fully compact the plasticity is permitted to spread in from the compression face of the section. Further, if the compression flange is fully compact and there is a tongue plate next to the compression edge of a semi-compact or slender web, the distance the plasticity may spread in from the compression face is limited (E.2). If the web is semi-compact, the plasticity must not extend beyond the compression edge of the web. If the web is slender, $p = p_m$ for the web.

When M_u is acting, the spread of plasticity will almost certainly be increased.

4.4 Assumed section (4.5.2.3)

The definitions of the effective section, net effective section and the net section are given in the section entitled 'Definitions'. For the effective section and the net effective section, the reduced thicknesses are taken as follows:

1) non-slender elements with HAZ effects: the thickness $k_z t$ is determined in the softened parts. The extent of the softening and k_z are as given in Sections 4.4.3 and 4.4.2,
2) slender elements with HAZ effects: the reduced thickness in the softened part is the lesser of $k_z t$ and $k_L t$. In the rest of the element it is $k_L t$ and
3) slender elements without HAZ effects: the thickness $k_L t$ is used for the whole element, with k_L being found as in Section 4.3.4.

The deduction for a hole located in a reduced thickness region is based on the reduced thickness. k_L for a reinforced element is applied to the basic plate thickness and to the area of the reinforcement.

For a slender flange element lying nearer to the neutral axis than the extreme fibre tensile material, k_L in Figure 4.5 can be modified by using $\epsilon = ([250y_1]/[p_o y_2])^{0.5}$ if the element is mainly parallel to the axis of bending.

Reinforcement in the form of an outwardly facing lip that forms part or all of the compression flange of a slender section should be ignored when determining the moment resistance.

The assumed thickness for a welded element in a semi-compact or a slender section is as follows:

1) the effects of HAZ softening in the material, less than $k_z y_1$ from the elastic neutral axis of the gross section, is ignored,
2) where $y > k_z y_1$ from the neutral axis, for HAZ material, k_z is replaced by $k_{zy} = (k_z + 1 - [y/y_1])$.

4.5 Hybrid sections and semi-compact sections

For a hybrid section containing different strength parent metals, the lowest value of p_o within the section may be used to determine the hybrid section's moment capacity (4.5.2.4). Alternatively, each element may be classified by its p_o value, then:

1) using normal plastic bending theory, for a fully compact section M_{RS} is found using the p_o for each element, and for welded members using the net effective section, but
2) for other sections,

unwelded,	semi-compact	$M_{RS} = (p_o Z_n)/(\gamma_m)$
	slender, the smaller of	$M_{RS} = (p_o Z_e)/(\gamma_m)$ or $(p_o Z_n)/(\gamma_m)$
welded,	semi-compact	$M_{RS} = (p_o Z_{ne})/(\gamma_m)$

but using the values of p_o and Z at the point in the section that gives the lowest values of M_{RS} (4.5.2.4).

For semi-compact sections, M_{RS} can be found as follows (4.5.2.5):

$$M_{RS} = M_s + ([\beta_o - \beta]/(\beta_o - \beta_1]) \times (M_f - M_s)$$

where:

β is the β value for the most critical element in the section,
β_o is the semi-compact limiting value of β for the same element,
β_1 is the fully compact limiting value of β for the same element,
M_f is the fully compact value of M_{RS} and
M_s is the semi-compact value of M_{RS}.

4.6 Shear force resistance (4.5.3)

4.6.1 Section classification

Regarding resistance to shear force, a compact section is unaffected by buckling, while a slender section needs to be checked for buckling (4.5.3.1). The sections are classified as shown.

1) For sections without tongue plates, but with shear webs aligned in the plane of the loading, the section is compact if $(d/t) \leq 49\epsilon$ and slender if $(d/t) > 49\epsilon$, where:
   - d is the clear depth of the web between the flanges or for inclined webs, measured on the slope,
   - p_o is the limiting stress for bending and overall yielding (Tables 4.1 and 4.2),
   - p_v is the limiting stress in shear (Tables 4.1 and 4.2) and
   - ϵ $= (250/p_o)^{0.5}$ which is equal to $(150/p_v)^{0.5}$ (as these two equations do not always give the same answer, $(150/p_v)^{0.5}$ is preferred) and
   - t is the web thickness.
2) For sections having shear webs with tongue plates that comply with Section 5.4.5, the shear force resistance can be found using Sections 5.4.3.1 to 5.4.3.5, provided the v-factors are as follows:
   - v_l is the elastic critical shear buckling factor determined as in Section 5.4.3.3 and
   - v_{tf} is the tension field factor, which in this case is zero.
3) Solid bar: compact.
4) Round tube is classified as for axial compression (4.3.2.5 and 4.3.3.4b), where $\beta = 3(D/t)^{0.5}$. If $\beta \leq \beta_o$, the round tube is compact. If $\beta > \beta_o$, the round tube is slender. The limiting values of β_o are given in Table 4.3.

4.6.2 Compact sections (4.5.3.2)

If there is no moment, V_{RS}, the factored shear force resistance, is as follows:

$$V_{RS} = (p_v A_v)/(\gamma_m)$$

where A_v is the effective shear area, p_v is the limiting stress found from Tables 4.1 and 4.2 and γ_m is the material factor found from Table 3.3.

The value of A_v is found as follows:

1) If there is no HAZ softening and the section has shear webs without tongue plates, then

$A_v = 0.8\ NDt$

where D is the overall depth of the web measured to the outer surfaces of the flanges, N is the number of webs and t is the web thickness.

Provided any small holes do not occupy more than 20 per cent of the clear web depth, between the flanges, they may be ignored.

2) If the webs in 1) above are affected by HAZ softening, then

$A_v = N(0.8Dt - [\{1 - k_z\}d_z t])$

where d_z represents the total depth of HAZ material within the clear depth of the web between the flanges (4.4.3) and k_z is the softening factor (4.4.2).

If the web is continuously welded longitudinally at any point in its depth or is welded over its full depth, V_{RS} equals k_z times the unwelded value.

3) $A_v = 0.8A$ or $0.8A_e$ for a solid bar

4) $A_v = 0.6A$ or $0.6A_e$ for a compact round tube
where A is the section area if there is no HAZ softening. If there is HAZ softening, A_e is the effective section area, which equals k_z times the true HAZ material thickness.

For sections containing shear webs, V_{RS} is found as for plate girders (5.4.3.2 and 5.4.3.5).

4.6.3 Slender sections (4.5.3.3)

For sections with slender shear webs, not having tongue plates and orientated in the plane of loading, V_{RS} in the absence of moment is the lower of either:

1) for the yielding check, 1) and 2) of the previous section 4.6.2 (Compact sections), as appropriate, is used (4.5.3.2a or 4.5.3.2b) or
2) for the buckling check
V_{RS} (in kN) = $(340\ Nt^3)/(d\gamma_m)$.

4.6.4 Inclined shear webs (4.5.3.4)

For compact sections, Sections 4.5.3.2a and 4.5.3.2b should be used for inclined shear webs, provided D is measured normal to the neutral axis (4.5.3.4).

For slender inclined webs (4.5.3.3),

$V_{RS} = (\text{Cos } \theta[\{340\ Nt^3\}/\{d\gamma_m\}])$,

where θ is the angle between the plane of the applied loading and the web.

4.6.5 Tongue plates (4.5.3.5)

For sections containing shear webs that have tongue plates complying with Section 5.4.5, V_{RS} is found using Sections 5.4.3.1 to 5.4.3.5. v_1 is determined using Section 5.4.3.3 and $v_{tf} = 0$.

4.7 Combined moment and shear forces (4.5.4)

If V, the shear force under factored loading, is less than $0.5V_{RS}$, M_{RS} is unaffected by coincident shear. V_{RS} is calculated in accordance with Section 4.5.3.3 (4.5.4.1).

If $V > 0.5V_{RS}$, then M_{RS} is modified to M_{RSO}, where for sections with shear webs, connected to flanges at both longitudinal edges (4.5.4.2):

$M_{RSO} = M_{RS}(1 + [1 - \alpha][0.6 - \{\langle 1.2V\rangle/\langle V_{RS}\rangle\}])$ and with

$M_{RSO} = M_{RS}(1.6 - [\{1.2V\}/\{V_{RS}\}])$ for all other sections.

α is the ratio of minimum to maximum shear stress in the web assuming an elastic stress distribution.

If the section is classified for bending as slender, or is affected by HAZ softening, α is based on the assumed section (4.5.2.3).

4.8 Web bearing (4.5.5)

Web bearing considers webs subjected to localized forces, such as concentrated loads or reactions applied to beams (4.5.5.1).

For unstiffened webs, where the web itself will carry the localized force without a bearing stiffener, both the following requirements must be met (4.5.5.2):

1) $p_{w1} \leq (p_a/\gamma_m)$ or $(k_z p_a \gamma_m)$ and
2) $p_{w2} \leq (p_s/\gamma_m)$

where:

p_{w1} is the stress arising at the extreme edge, assuming a 45° angle of dispersion either side of the localized force,
p_{w2} is the stress arising at the mid-point, again assuming a 45° angle of dispersion either side of the localized force,
p_s is the buckling stress for the web treated as a thin column between the flanges and
k_z is the softening factor for the HAZ material.

The expression $p_{w1} = (k_z p_a \gamma_m)$ is only used when the web is welded to the flange and HAZ softening occurs.

The value of p_s is read in accordance with Section 4.7.4.1 from Figure 4.10a, using the curve that intercepts the stress axis at a value of p_o as found from Tables 4.1 and 4.2. λ, the slenderness parameter used in Figure 4.10a, must be at least or greater than 3.5d/t and should allow for the relative lateral movement of the flanges as the web buckles (4.7.4.1 and 4.5.5.2).

When a tongue plate is used the equations $p_{w1} \leq (p_a/\gamma_m)$ or $(k_z p_a \gamma_m)$ must be satisfied at both the upper edge of the thin web and at the top edge of the tongue plate (4.5.5.3).

Bearing stiffeners are designed as compact sections. They can be designed by assuming they resist the entire bearing force, unaided by the web, with the stiffener being checked as a strut for local squashing, out-of-plane column buckling and any bending effects (4.7 and 4.8). Alternatively, the bearing stiffener may be designed, more economically, as a plate girder stiffener using Section 5.4.5 (4.5.5.4).

4.9 Lateral torsional buckling (4.5.6)

With certain exemptions, a beam should be checked for lateral torsional buckling. The exceptions are (4.5.6.1):

1) a beam supported throughout its length against lateral movement,
2) bending about the minor axis and
3) when lateral supports are provided to the compression flange at spacings not greater than $40\epsilon r_y$. $\epsilon = (250/p_o)^{0.5}$ and r_y is the minor axis radius of gyration of the section (4.5.6.2).

The condition $M \leq M_{Rx}$ must be satisfied in every unsupported bay between points of lateral support (4.5.6.3), where

M is the moment caused by the factored loading in the length being considered,

M_{Rx} $= p_sS/\gamma_m$ and is the factored moment of resistance to lateral torsional buckling and

S is the plastic section modulus of the gross section, unreduced for HAZ softening, holes or local buckling.

p_s is the buckling stress and is read from Table 4.9 (4.5.6.5).

The value of M may be assumed as the maximum value arising in the bay being considered, or it may be assumed that $M = \bar{M}$. For a moment with a linear variation in the length being considered, then (4.5.6.4):

1) for $1.0 > (M_2/M_1) \geq -0.5$, $\bar{M} = (0.6M_1 + 0.4M_2)$ and
2) for $(M_2/M_1) < -0.5$, then $\bar{M} = 0.4M_1$
   where M_1 is the maximum factored moment and M_2 is the minimum factored moment as shown in Figure 4.8. Other forms of moment variation are considered in Appendix H (4.5.6.4).

4.9.1 Buckling stress (4.5.6.5)

p_s is found from Figure 4.9 when $\lambda \leq 130$, by using the curve that intercepts the stress axis at a stress of p_l, where

$p_l = p_o$ for unwelded fully compact sections and
$p_l = ([\gamma_m M_{RS}]/S)$ for all other sections including hybrid sections.

For $\lambda > 130$, Appendix K, Figure K1 is used (4.5.6.5).

The value of M_{RS} is found as described in Section 4.5.2, allowing for HAZ softening and local buckling, but making no deductions for holes. Where HAZ softening has taken place, M_{RS} is found for the most unfavourable section in the bay being considered, even if the softening occurs only locally. If the HAZ softening occurs only at the ends of the bay and does not extend along the member more than the section width, it may be ignored when considering lateral torsional buckling (4.5.6.8).

4.9.2 Slenderness parameter λ (4.5.6.6)

λ is the lateral torsional buckling slenderness parameter used in Figure 4.9. Any one of the three following equations may be used (4.5.6.6):

$$\lambda = \lambda_y = (l/r_y)$$

where r_y is the gross section minor axis radius of gyration and l is the effective length for lateral torsional buckling. l is the distance between points of lateral support, but is found using Appendix H for the following four cases:

1) for a beam with effective lateral restraints at its ends only, l is found from Table H.1, but if the effective lateral restraints are at intervals along the beam length, l is the length between the restraints (H.1.1),
2) for cantilever beams with no intermediate restraints and with no moment applied at the tip, l is found from Table H.2. If there are intermediate restraints, or there is a moment at the tip, l is found using 1) above (H.1.1) and 3) below (H.1.2),
3) for beams subjected to loading between points of lateral support, when for example the load is applied to a compression flange and both the flange and the load are free to deflect laterally, relative to the beam centroid, l is increased as given in Table H.1. If the beam has effective lateral restraints at intervals, then l is 1.2 times the distance between the restraints (H.1.2) and
4) for beams subjected to normal loading, where the compression flange is laterally unrestrained with both flanges free to rotate in plan, and the bearing of the bottom flange on the supports provide the only torsional restraint (4.5.6.6).

The general expression for λ is as follows:

$$\lambda = \pi(ES/M_{cr})^{0.5}$$

where E is the modulus of elasticity and M_{cr} the elastic critical uniform moment. Appendix H gives values for M_{cr} for a doubly symmetrical uniform section and for a uniform section symmetrical about the minor axis only, assuming a uniform moment in both cases (H.2). Beams of varying section throughout their length are discussed in Section H.3.

For the channel and I-section members shown in Table 4.7, λ must not exceed $\lambda_{y.}$ λ is determined as follows:

$$\lambda = ([X\lambda_y]/[1 + Y\{\lambda_y/\langle D/t_2\rangle\}^2]^{0.25})$$

where D is the overall section depth, t_2 is the flange thickness, X = 1 or the value given in Table 4.7 and Y = 0.05 or the value given in Table 4.7.

λ can also be found using the above equation and Table 4.7, for I-beams and channels that are not exactly the same as those given in Table 4.7. X and Y are as for the equivalent simple lip having the same internal depth C, but the section with the actual reinforcement is used to determine λ_y (4.5.6.6).

To provide effective lateral restraint, it is assumed that the bracing system points of restraint in any one span resist a total lateral force, exerted by the compression flange under factored loading, of 3 per cent of the compression in the flange (4.5.6.7). Two or more parallel beams must be restrained either by a triangulated bracing system, or by tying them to a robust independent support. For more than three parallel beams, the sum of the lateral forces to be resisted is found from the three largest compressive forces (4.5.6.7).

5 Static design of tension members (4.6)

5.1 Introduction

The design of axially loaded tension members is usually straightforward in that

$$P \leq P_{RS}$$

where P is the axial tensile force due to factored loading and P_{RS} is the factored tensile axial resistance.

If the tension member contains material affected by HAZ, then the material strength is reduced.

If the tension member has eccentric end connections, the interaction between the axial load and the moments introduced by the eccentric end connections must be considered (4.8). A simplified procedure is available for certain cases (4.6.3).

5.2 Tension resistance (4.6.2)

P_{RS}, the factored tensile axial resistance, is the lesser of either:

a) the general yielding along the member (4.6.2.2) or
b) the local failure at a critical section (4.6.2.3).

For general yielding, P_{RS} is based on the general cross-section of the member along its length, ignoring the effect of end connections, localized HAZ regions and occasional holes (4.6.2.2).

1) For a member with only localized HAZ softening or free from HAZ softening,

$$P_{RS} = ([p_o A]/[\gamma_m]).$$

2) If the member has HAZ material generally along its length, then

$$P_{RS} = ([p_o A_e]/[\gamma_m]).$$

A_e is the effective section area and is k_z times the true area of any softened zone. k_z is found from Section 4.4.2. The extent of the zone is found from Section 4.4.3.

For local failure, P_{RS} is based on the most critical of the following (4.6.2.3):

1) $P_{RS} = ([p_a A_n]/[\gamma_m])$ for a section free from HAZ softening and
2) $P_{RS} = ([p_a A_{ne}]/[\gamma_m])$ for a section containing HAZ material,

where A_n is the net section area, with a deduction for holes and A_{ne} is the net effective section area.

A_{ne} is found in the same way as A_e, but with a deduction for any holes. For holes in the HAZ regions, the deductions are based on $k_z t$ (4.4.2).

For staggered holes, A_n and A_{ne} are calculated for the least favourable section. For a diagonal or zig-zag section:

$$A_n = A - H \text{ or}$$
$$A_{ne} = A_e - H,$$

where $H = \Sigma A_h - \Sigma([x^2 t]/[4y])$. x is the longitudinal pitch of the holes, y is the transverse pitch of the holes and Σ_h is the sum of the hole areas on the diagonal or zig-zag section being considered (4.6.2.4). The lower of the values of A_n and of A_{ne} are then used in the local failure equations for P_{RS} above.

For hybrid sections with materials of different strengths, the tension capacity is found by summing the resistance of the various parts as found in the local failure equations of Section 4.6.2.3 for P_{RS} above (4.6.2.5).

5.3 Eccentrically connected ties (4.6.3)

Angles connected through one leg, flange-connected tees and web-connected channels are considered as eccentrically connected ties and, if they are single bay tension members, may be designed as axially loaded. The variation in the stress in the outstanding leg or legs can be ignored, provided, when determining A_n or A_{ne} for the local failure check (4.6.2.3), part of the outstanding leg area is also deducted from the gross area. This deduction is in addition to the deduction for HAZ effects and the deduction for holes. The amount of the outstanding leg area to be deducted is:

1) $0.6A_o$ for a single component connected to one side of a gusset or
2) $0.2A_o$ for a double component symmetrically connected either side of a gusset.

A_o is the effective area of the outstanding leg or legs lying clear of the connected element. The fillet is ignored.

If the eccentrically connected ties are continuous over a number of bays, only the outer ends of the end bays need be considered using Section 4.6.3. For other areas of the bays, Section 4.6.2.3 is used, without any deduction for outstanding legs. General yielding should also be checked in accordance with Section 4.6.2.2.

5.4 Design example 1

Determine the factored tension resistance P_{RS} of a channel section 200 mm × 80 mm, made from 6063-T6. The section is free from any effects of end connections, localized HAZ regions and occasional holes (4.6.2).

Table 4 of BS 1161 gives $A = 33.9 \text{ mm}^2 \times 10^2$.
For 6063-T6, Table 4.1 of BS 8118 gives $p_o = 160 \text{ N/mm}^2$ and $p_a = 175 \text{ N/mm}^2$.
Table 3.3 gives $\gamma_m = 1.2$.

P_{RS} is the lesser value of either:

a) general yielding along the member (4.6.2.2a) or
b) local failure at a critical section (4.6.2.3a).

For general yielding, $P_{RS} = p_o A/\gamma_m$
$= (160\ N/mm^2 \times 33.9\ mm^2 \times 10^2)/1.2$
$= 452\ kN.$

For local failure (here, $A_n = A$), $P_{RS} = p_a A_n/\gamma_m$
$= (175\ N/mm^2 \times 33.9\ mm^2 \times 10^2)/1.2$
$= 494.4\ kN.$

Giving $P_{RS} = 452\ kN.$

5.5 Design example 2

Determine the factored tension resistance P_{RS} of an equal bulb angle section 60 mm × 60 mm, made from 6063-T6. The section is free from any effects of end connections, localized HAZ regions and occasional holes (4.6.2). The angle is connected through one leg only and is continuous over several bays (4.6.3).

Table 7 of BS 1161 gives $A = 494\ mm^2$ and $t = 3$ mm.
For 6063-T6 extrusion, Table 4.1 of BS 8118 gives $p_o = 160\ N/mm^2$ and $p_a = 175\ N/mm^2$. Table 3.3 gives $\gamma_m = 1.2$.

P_{RS} is the lesser value of either:

a) general yielding (4.6.2.2a) or
b) local failure at a critical section (4.6.2.3a).

For general yielding:

$P_{RS} = p_o A/\gamma_m$
$= (160\ N/mm^2 \times 494\ mm^2)/1.2$
$= 65.87\ kN.$

For local failure, as the angle is connected through one leg only and is continuous over several bays, two calculations are required (4.6.3).

For the inner bays, (here, $A_n = A$):

$P_{RS} = p_a A_n/\gamma_m$
$= 175\ N/mm^2 \times 494\ mm^2/1.2$
$= 72.04\ kN.$

For the outer ends of the end bays, A_n is reduced by having part of the outstand leg area deducted from the gross area. The amount to be deducted is $0.6A_o$. $A_o = 0.5(A - 0.5[3t]^2) - 0.5t^2$ where $0.5(3t)^2$ is the fillet area and $0.5t^2$ is the common corner area.

$$
\begin{aligned}
A_o &= 0.5(A - 0.5[3t]^2) - 0.5t^2 \\
&= 0.5(494 \text{ mm}^2 - 0.5[3 \times 3 \text{ mm}]^2) - 0.5 \times (3 \text{ mm})^2 \\
&= 222.25 \text{ mm}^2 \text{ and} \\
A_n &= A - 0.6A_o \\
&= 494 \text{ mm}^2 - 0.6 \times 225.25 \text{ mm}^2 \\
&= 358.85 \text{ mm}^2
\end{aligned}
$$

$$
\begin{aligned}
P_{RS} &= (175 \text{ N/mm}^2 \times 358.85 \text{ mm}^2 \times 10^2)/1.2 \\
&= 52.33 \text{ kN.}
\end{aligned}
$$

For the end angle sections, $P_{RS} = 52.33$ kN and is for local failure. For the internal bays, $P_{RS} = 65.87$ kN and is for general yielding.

6 Static design of compression members

6.1 General

For compression members care must be taken over the effects of any high residual stress, any initial deviation from straightness, of local buckling of thin members, of welding, and the effect of any holes.

Compression members must be classified at each cross-section, at the least favourable position, as compact or slender (4.7.2 and 4.3.3). A compact section is one that has no premature local buckling, while a slender section is one where the local buckling of any thin elements reduces its resistance. HAZ softening also reduces the member's resistance.

The section classification for axial compression is considered in Chapter 3 above, 'Static design of members', under sub-headings 3.5, 3.5.1, 3.5.2 and 3.5.3 (4.3.3).

For axially loaded compression members or struts, the following checks are made (4.7.1.1):

1) a flexural buckling check on the overall buckling of the member as a whole is always made on a column (4.7.3 and 4.7.4),
2) a torsional buckling check on the overall buckling of the member as a whole is generally, but not always, required (4.7.3 and 4.7.5) and
3) a local squashing check on the weakest cross-section along the length of a strut is only required where there is significant local

weakening due to holes or welding and when the strut has a low slenderness ratio.

For combined axial loading and bending, Section 4.8 should be used, but, if the strut has moments introduced by eccentric end connections, it may be possible to use the simplified procedure of Section 4.7.9 (4.7.1.2).

For overall buckling, using 1) and 2) above:

$$P \leq P_R, \quad \text{where } P_R = (p_s A)/(\gamma_m).$$

P is the axial compressive force due to factored loading and
P_R is the factored axial resistance based on the overall buckling.

Consideration must be given to column buckling failure about both principal axes and the lower value of p_s used. The value of p_s to be used when $\lambda > 130$ is found from Appendix K.

6.2 Column buckling (4.7.4)

The column buckling stress p_s is found from Figure 4.10, in accordance with Section 4.7.6 (4.7.4.1). λ is the column buckling slenderness parameter as follows:

$$\lambda = l/r$$

where l is the effective length and r is the radius of gyration for the appropriate direction of buckling. The effective length is:

$$l = KL$$

where L is the length between points of lateral support, or, for a cantilever strut, its length, and K is the effective length factor for struts, found from either Table 4.8, which gives K a value in the range from 0.7 to 2.0, or from a knowledge of the end conditions (4.7.4.2). If, at a directionally restrained end, there is HAZ softening that substantially or wholly affects the restraint, then that restraint is ignored. The gross section is used to determine r (4.7.5.1).

6.3 Torsional buckling (4.7.5)

The torsional buckling of closed hollow sections, doubly symmetrical I-sections and sections composed entirely of radiating outstands as in angles, cruciforms and tees, are classified as compact (4.3.3) and can be ignored (4.7.5.1).

λ is based on the gross area of the section and is found using a) or b) below or Appendix J. Appendix J presents a general procedure for determining λ for any section, including those in Table 4.9. The simplified procedure given in b) below uses empirical formulae for the limited range of shapes given in Table 4.9 (4.7.5.2b).

The general formula is:

a) $\lambda = \pi([EA]/[P_{cr}])^{0.5}$

but for the sections in Table 4.9,

b) $\lambda = k\lambda_t$

$\lambda_t = \lambda_o$ for angles, cruciforms and tees, but for channels and top-hats

$\lambda_t = (\lambda_o)/(1 + [\{Y\lambda_o{}^2\}/\{\lambda_x{}^2\}])^{0.5}$

where s, X, Y and λ_o are the torsional buckling parameters for struts, found from Table 4.9; s and X are required for Figure 4.11. λ_x is the effective slenderness for column buckling about the x–x axis as defined in Table 4.9 (4.7.5.2). The strut torsional buckling interaction factor k is found from Figure 4.11. The value of p_s, for torsional buckling, is found from Figure 4.12 (4.7.5.3).

6.4 Strut curve selection (4.7.6)

Table 4.10 is used to determine whether p_s is to be read from Figures 4.10a, b and c for column buckling or from Figures 4.12a and b for torsional buckling. For column buckling it is necessary to determine whether the strut is welded or unwelded and whether it is symmetric, mildly asymmetric or severely asymmetric. For torsional buckling it is necessary to decide whether or not the section is composed of radiating outstands. In using Figures 4.10 a, b or c and Figures 4.12 a or b, the correct curve is the curve meeting the stress axis at a stress p_l, determined as in Section 4.7.6.2 (4.7.6.1).

Except for sections composed of radiating outstands, p_l is found as follows (4.7.6.2):

$p_l = p_o$ for compact sections with no HAZ effects and
$p_l = (A_e/A)p_o$ for all other sections.

If the strut does not meet the tolerances of straightness and twist, laid down in BS 8118: Pt 2, for extruded members, then p_l is multiplied by s, where

$$s = 0.6 + 0.5 \exp(-0.02\lambda), \text{ but with } s \leq 1.0.$$

The effective section applies to strut sections that are classified as:

a) affected by HAZ softening,
b) slender or
c) both affected by HAZ softening and slender (4.7.6.3).

The effective section is obtained by using the least favourable cross-section, with reduced thickness and no deduction for holes, but it may be modified for welded members (4.7.6.5).

1) For unreinforced slender sections free from HAZ softening, the element thickness is k_L times t (4.3.4 and 4.7.6.3). For reinforced slender sections free from HAZ softening, the element thickness is k_L times the area of reinforcement plus k_L times the basic plate thickness.
2) For compact sections with HAZ softening, the softened zone thickness must be reduced to give an assumed area of k_z times the true area. k_z is found from Section 4.4.2 and the extent of the HAZ softening is found from Section 4.4.3.
3) For slender sections free from HAZ softening, 1) above is used to determine the reduced thickness.
4) For slender sections, with HAZ softening but not located in the slender element, then the reduced thickness is found as in 2) above.
5) For a slender section with HAZ softening, the reduced thickness in the softened part is either k_Lt or k_zt, whichever is the lesser, and k_Lt elsewhere.

Local buckling and torsional buckling are closely related in sections composed entirely of radiating outstands, e.g. angles, cruciforms and tees (4.7.6.4).

1) If the section contains only unreinforced outstands, then for torsional buckling p_s is found from Figure 4.12b and not Figure 4.12a. The relevant diagram for column buckling is not changed. To determine p_l for use with Figures 4.10 and 4.12, A_e is based on the effective section, where the normal reduction is made for HAZ softened zones, but with no reduction for local buckling, that is $k_L = 1$. For such a section free of HAZ effects, $p_l = p_o$ (4.7.6.4).

2) If the section contains outstands with tip reinforcement, such that the reinforced element buckles as a unit (mode 1) taking the reinforcement with it (4.3.2.3), then the procedure described in 1) above is used. If, however, the sub-elements comprising the reinforced elements themselves buckle as individual elements (mode 2) with the junctions between them staying straight (4.3.2.3), then Figure 4.12a is used and the effective section calculated as in Section 4.7.6.3 (4.7.6.4).

For struts containing localized welds, the value of p_l, used for strut curve selection, is based on the most unfavourable section even if the HAZ effects are localized or as a result of the welding-on of temporary attachments. If the HAZ softening position is specified and does not extend longitudinally more than the least overall width of the member, its presence may be ignored when considering overall buckling, provided it is positioned at zero or near zero curvature in the buckled form of the strut. However, a local squashing check may be needed.

6.5 Local squashing (4.7.7)

For local squashing, $P_{RS} \geq P$ for the most unfavourable section along the strut length, where P is the axial thrust under factored loading. P_{RS} is found as follows:

1) $P_{RS} = ([p_a A_n]/\gamma_m)$ for compact sections unaffected by HAZ effects and
2) $P_{RS} = ([p_a A_{ne}]/\gamma_m)$ for all other sections.

A_{ne} is A_e minus the deduction for unfilled holes, where A_e is as used when considering overall buckling (4.7.6.3 and 4.7.6.4). If the holes are in a reduced thickness area of the member, the deductions are based on the reduced thickness (4.7.7).

6.6 Hybrid sections (4.7.8)

For struts made up of more than one parent metal, it is necessary for each element to be classified according to its own p_o. The overall column or torsional buckling resistance P_R is found by assuming a single value of p_o equal to the weighted average, according to the gross areas, of the p_o values for the parts under consideration. The total resistance to local squashing, P_{RS}, is the sum of the resistance of the various parts.

6.7 Certain types of eccentrically connected struts (4.7.9)

Section 4.8 gives an interaction procedure for eccentrically connected struts, but a simpler method may be used if there is no designed applied bending and the attachments are sufficient to prevent rotation in the plane of the connected element. The simpler method may be used on the following:

1) back-to-back angles connected on one side of a gusset,
2) a single angle connected through one leg only,
3) a single channel connected by its web only and
4) a single tee connected by its table only.

When checking 1) to 4) above for column buckling that is out of the plane of the attached element or elements, the eccentricity of the loading is ignored and a reduced axial compression resistance is used. The reduced axial compression resistance is taken as 40 per cent of that found for centroidal loading, using the radius of gyration about the axis parallel to the gusset. The eccentricity of the loading is assumed not to affect the torsional buckling resistance (4.7.9.1).

Where channels, double angles or tees are constructed back-to-back and connected either side of end gussets, they are designed as centroidally loaded monolithic members. This design approach is valid only when the two components are connected securely together at their ends and are connected at their third points, using spacers equal to the gusset thickness (4.7.9.2).

6.8 Battened struts (4.7.10)

Sections 4.7.3 to 4.7.7 inclusive cannot generally be applied to battened struts, but the resistance of certain battened struts, regarded as monolithic members, can be found in the normal way, provided they satisfy the following (4.7.10.1):

1) they are axially loaded,
2) they have two main components, joined by equally spaced battens, with the cross-section symmetrical about an axis normal to the battens,
3) the battens should be in pairs, but single battens are allowed if the main components are toe-to-toe angles or tees,
4) $\lambda_2 \leq 0.8\lambda_1$, where λ_1 is the slenderness parameter for column buckling of the complete member about an axis parallel to the

battens and λ_2 is the slenderness parameter for column buckling of the complete member about an axis normal to the battens,

5) $\lambda_3 \leq 0.7\lambda_2$, where λ_3 is the slenderness parameter for the buckling of one of the main components between the battens, based on the most critical of either column or torsional buckling,
6) the design of the batten system should be such as to resist a total shear force V, in the plane of the battens, equal to 2.5 per cent of the axial force in the whole member under factored loading and
7) each batten connection to each main component is designed to transmit the following simultaneous actions under factored loading: a) a longitudinal shear of (Vd/Na) and b) a moment of (Vd/2N), acting in the plane of the batten where:
   a is the spacing of the main components measured to the centroids of the connections to each batten,
   d is the longitudinal spacing between the centres of the battens and
   N is the number of battens at each position (1 or 2).

The weakening effects of HAZ softening and local buckling must also be considered (4.7.10.1).

6.9 Design example 1

An extruded I-section, 160 mm × 80 mm in 6082-T6 material is axially loaded without bending. The distance between the two ends is 2 000 mm and both ends are effectively held in position and restrained in direction. Determine the maximum factored axial compressive resistance P_R, ignoring any bolted or welded connections.

For compression members, a check on flexural buckling, torsional buckling and local squashing are required (4.7.1.1).

The section must be classified as either compact, where local buckling can be ignored, or slender, where local buckling lowers the moment resistance, using Section 4.3.3 (4.7.2).

The section is compact if $\beta \leq \beta_o$, and slender if $\beta > \beta_o$ (4.3.3.4). The sizes of the I-section are given in Table 5 of BS 1161 and shown in Figure 6.1 opposite. For an element under uniform compression, using Section 4.3.2.2 and Figure 4.1 of BS 8118:

a) for the unreinforced flange outstand element,
   $\beta = b/t = (40\text{ mm} - [2 \times 7\text{ mm}])/11\text{ mm} = 2.364$,
b) for the unreinforced web internal element,
   $\beta = b/t = (160\text{ mm} - [2 \times 11\text{ mm}] - [3 \times 7\text{ mm}])/7\text{ mm} = 16.71$.

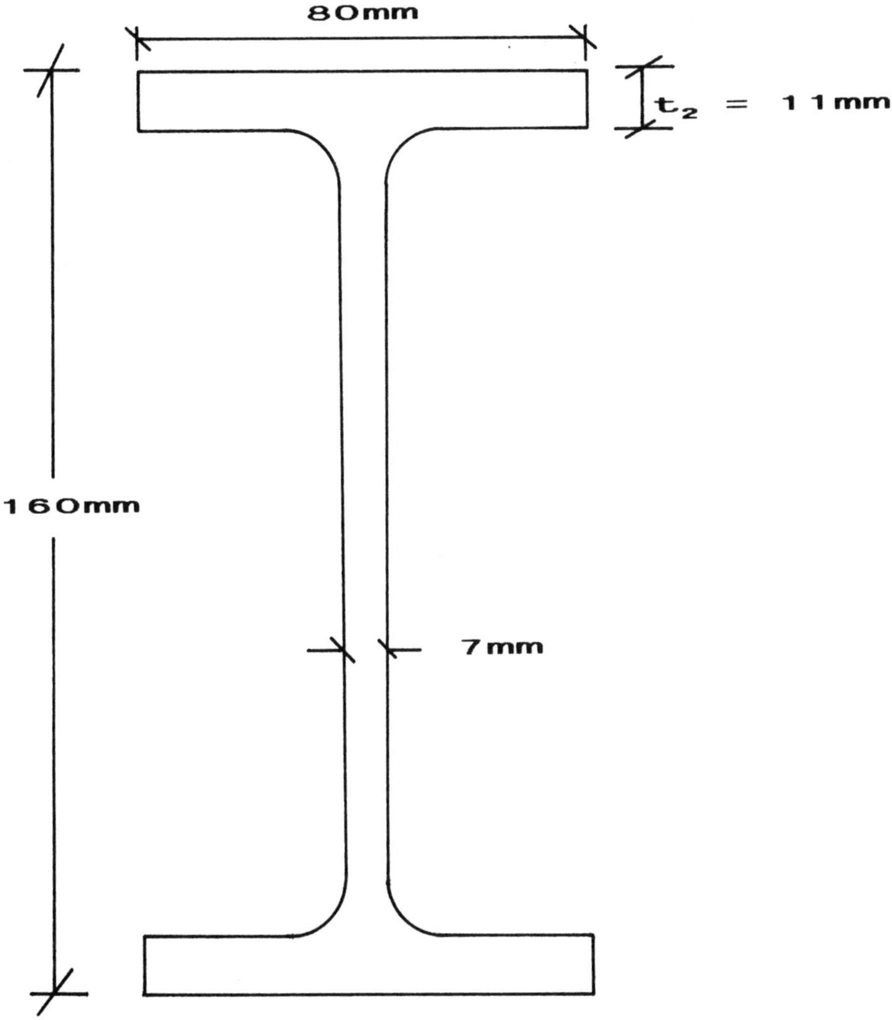

Figure 6.1 Section through the I-section

For the unwelded outstand flange element, $\beta_o = 7\epsilon$, where $\epsilon = (250/p_o)^{0.5}$. From Table 4.1, $p_o = 255$ N/mm^2 giving $\epsilon = 0.99$ and $\beta_o = 6.93$. Thus β (2.364) $< \beta_o$ (6.93), the outstand element is compact and local buckling can be ignored (4.3.3.2).

For the unwelded internal web element, $\beta_o = 22\epsilon$, giving $\beta_o = 21.78$.

Thus β (16.71) < β_o (21.78), the internal element is compact and local buckling can be ignored (4.3.3.2).

The I-section is compact.

The factored resistance to overall buckling is:

$P_R = p_s A/\gamma_m$ (4.7.3),

where:
$A = 2\,820\ mm^2$ (Table 5, BS 1161),
$\gamma = 1.2$ (Table 3.3) and
p_s is found for column buckling about both principal axes and the lower value taken, using Figure 4.10 (4.7.4.1).

$l = KL$ (4.7.4.2), where $K = 0.7$ (Table 4.8) and $L = 2\,000$ mm, giving $l = 1\,400$ mm. The column buckling slenderness parameter $\lambda = l/r$, where r is the lower of r_x (64.5 mm) and r_y (18.3) (Table 5, BS 1161), giving $\lambda = 1\,400\ mm/64.5\ mm = 21.71$ and $\lambda = 1\,400\ mm/18.3\ mm = 76.5$. Table 4.10 shows that Figure 4.10a is to be used for column buckling (symmetric, unwelded strut).

We now use Section 4.7.6 for the strut curve selection. As the section is compact with no HAZ effects, $p_l = p_o$ (4.7.6.2) for 6082-T6 material, giving $p_l = p_o = 255\ N/mm^2$. Using the vertical y axis of Figure 4.10a, go up to $p_s = 255\ N/mm^2$ and interpolate between the two curves $p_s = 240\ N/mm^2$ and $p_s = 260\ N/mm^2$. Follow the interpolated curve until it intercepts the $\lambda = 76.5$ line, giving $p_s = 98\ N/mm^2$.

$$P_R = (98\ N/mm^2 \times 2\,820\ mm^2)/1.2 = 230.3\ kN.$$

Torsional buckling may be ignored as the I-section is doubly symmetric (4.7.5.1).

For a compact section, free from HAZ effects, for local squashing (4.7.7):

$P_{RS} = p_a A_n/\gamma_m$,

where:

$p_a = 275\ N/mm^2$,
$A_n = A = 2\,820\ mm^2$ and
$\gamma_m = 1.2$ giving
$P_{RS} = (275\ N/mm^2 \times 2\,820\ mm^2)/1.2$
$= 646.25$ kN.

For the strut being considered, overall buckling of the member as a whole is the limiting value giving $P_R = 230.3$ kN.

6.10 Design example 2

An equal angle section in alloy 6082-T6 is 100 mm × 100 mm × 6 mm thick as shown in Figure 6.2 below. The section is eccentrically loaded by a connection, at each end, through one leg. The distance between the two ends is 2 000 mm and both ends are effectively held in position and

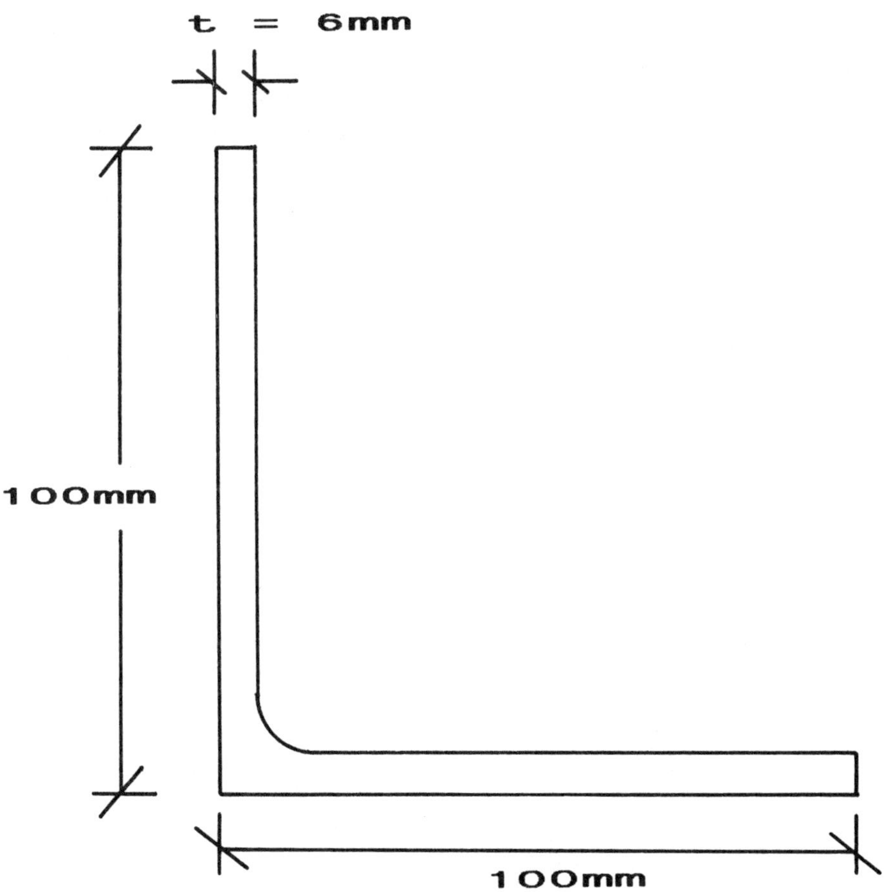

Figure 6.2 Equal angle section: thin range

restrained in direction. Determine the factored compressive axial resistance P_R. Any bolted or welded connections are to be ignored. There are no HAZ effects.

From Table 2 of BS 1161, $A = 1230\ mm^2$, $r_x = r_y = 30.7$ mm and $r_u = 38.5$ mm. Table 4.1 for 6082-T6 material gives $p_o = 255\ N/mm^2$ and $p_a = 275\ N/mm^2$ as the section thickness is less than 20 mm.

The member must be checked for flexural column buckling, torsional buckling and local squashing (4.7.1). If the strut was subjected to axial load plus bending then Section 4.8 would be used. The strut has eccentric end connections that introduce moments, and, in certain cases, the simplified procedure of Section 4.7.9 can be used rather than Section 4.8 (4.7.1.2).

Assuming rotation in the plane of the connected element is prevented and no deliberate bending is applied, as in this example, then for out-of-the-plane column buckling the eccentricity of loading can be ignored (4.7.9.1a). The axial compressive column buckling resistance becomes 40 per cent of the value obtained for centroidal loading using the radius of gyration about the axis parallel to the gusset. Torsional resistance is unaffected (4.7.9.1).

The section is classified in accordance with Section 4.3.3 as either compact ($\beta \leq \beta_o$), where local buckling can be ignored, or slender ($\beta > \beta_o$), where local buckling lowers the resistance (4.7.2).

For the unwelded outstand element, $\beta_o = 7\epsilon = 7(250/p_o)^{0.5}$ giving $\beta_o = 6.93$ (Table 4.3).

β is found using Section 4.3.2 and Figure 4.1 of BS 8118: Pt 1. For uniform compression $\beta = b/t$ (4.3.2.2) where $b = (100\ mm - [4 \times 6\ mm])$ and $t = 6$ mm (Table 2, BS 1161), giving $\beta = 12.67$. As β (12.67) $> \beta_o$ (6.93), the member is considered slender (4.3.3.4).

For column buckling (4.7.4) $\lambda = l/r$ (4.7.4.2), where $l = KL = 0.7$ (Table 4.8) $\times$ 2 000 mm = 1 400 mm. The value of r about the axis parallel to the connected leg is $r_x = 30.7$ mm giving $\lambda = 45.60$.

The resistance to overall buckling is $P_R = p_s A/\gamma_m$ (4.7.3), where $A = 1\ 230\ mm^2$ and $\gamma_m = 1.2$ (Table 3.3). p_s is found from Figure 4.10 using Section 4.7.6 (4.7.4.1).

The strut is unwelded, severely asymmetric (Note 2, Table 4.10) and, as $y_1/y_2 > 1.5$, Figure 4.10b is used. y_1/y_2 about the neutral axis of the angle = 73.7 mm/26.3 mm = 2.80.

To determine the appropriate curve, the value of p_l on the vertical stress axis is required (4.7.6.1). Considering Section 4.7.6.2, equal angle sections are composed entirely of radiating outstands, therefore local and torsional buckling are closely related (4.7.6.4).

If the section had not been composed of radiating outstands, then for other than compact sections with no HAZ effects $p_l = p_o(A_e/A) =$

$255\ N/mm^2(A_e/1\ 230\ mm^2)$ would have been used (4.7.6.2). It is assumed the member meets the tolerances of straightness and twist laid down for extruded material as given in BS 8118: Pt 2, A_e being found using Section 4.7.6.3 and there being no HAZ softening.

For buckling of slender sections, the true section is replaced by an effective section, which is found using a local buckling coefficient k_L (4.3.4.1). Using $\beta = 12.67$ and $\epsilon = 0.99$, for unwelded flat outstands, Figure 4.5, curve A (Table 4.4) gives $k_L = 0.634$.

The thickness of 6 mm being reduced to 6 mm $\times$ 0.634 = 3.81 mm, a reduction of 2.19 mm, giving $A_e = 1\ 230\ mm^2 - (2.19\ mm \times [100\ mm + 97.81\ mm]) = 796.80\ mm$, giving:

$$\begin{aligned} p_l &= (796.80\ mm/1\ 230\ mm) \times 255\ N/mm^2 \\ &= 165.19\ N/mm^2. \end{aligned}$$

Using $p_l = 165.19\ N/mm^2$ and $\lambda = 45.60$, Figure 4.10b gives $p_s = 123\ N/mm^2$. The resistance to overall buckling being:

$$\begin{aligned} P_R &= p_s A/\gamma_m \\ &= 123\ N/mm^2 \times 1\ 230\ mm^2/1.2 \\ &= 126.08\ kN. \end{aligned}$$

However, in this case $k_L = 1.0$ and $p_l = p_o = 255\ N/mm^2$ (4.7.6.4(2)). Using $p_l = 255\ N/mm^2$ and $\lambda = 45.60$, Figure 4.10b gives $p_s = 161\ N/mm^2$ and

$$\begin{aligned} P_R &= 0.4 p_s A/\gamma_m \\ &= 0.4 p_s \times 1\ 230\ mm^2/1.2 \\ &= 0.4 \times 161\ N/mm^2 \times 1\ 230\ mm^2/1.2 \\ &= 66.01\ kN. \end{aligned}$$

For torsional buckling, $P_R = p_s A/\gamma_m = p_s \times 1\ 230\ mm^2/1.2$ (4.7.3). As the member is classified as a slender equal angle section composed entirely of radiating outstands, torsional buckling must be considered (4.7.5.1).

λ is found using Table 4.9 (4.7.5.2b) where $\lambda = k\lambda_t$ (4.7.5.2b). As $\rho = R/t$ (Table 4.9) = 3t/t from Table 2 BS 1161, $\rho \leq 5$ and row 1 of Table 4.9 can be used.

For the angle section:

$$\begin{aligned} \lambda_t &= \lambda_o \\ &= \lambda_1 \end{aligned}$$

$= ([5 \times 100 \text{ mm}/6 \text{ mm}] - 0.6 \times 3^{1.5}[100 \text{ mm}/6 \text{ mm}]^{0.5})$
$= 70.61.$

To determine k, Figure 4.11 is used. $\lambda_u = l/r$ about the u axis, where $r = 38.5$ mm (Table 2, BS 1161) and $l = 1\,400$ mm, giving $\lambda_u = 36.36$, with $s = \lambda_u/\lambda_o = 36.36/70.61 = 0.52$ and $X = 0.6$ (Table 4.9 of BS 8118: Pt 1).

Using Figure 4.11 of BS 8118: Pt 1 gives k = 1.06, thus:

$$\lambda = k\lambda_t$$
$$= 1.06 \times 70.61$$
$$= 74.85.$$

The value of p_s for torsional buckling is found from Figure 4.12 (4.7.5.3). Table 4.10 shows that for torsional buckling generally, Figure 4.12a is used, but as the section is an angle composed entirely of radiating outstands Figure 4.12b is used (4.7.6.1). Further, $p_l = (A_e/A)p_o$ is not used, as $k_L = 1.0$ and $p_l = p_o = 255$ N/mm², as found for column buckling (4.7.6.4(2)).

Using Figure 4.12b, with $\lambda = 74.85$ and the curve for 255 N/mm², gives an intercept of $p_s = 104$ N/mm².

$$P_R = P_s A/\gamma_m$$
$$= 104 \text{ N/mm}^2 \times 1\,230 \text{ mm}^2/1.2$$
$$= 106.60 \text{ kN}.$$

The local squashing check would give $P_{RS} = p_a A_{ne}/\gamma_m$ (4.7.7). A_{ne} is the net effective section area of the most unfavourable section along the strut length. This occurs on one leg due to the loading being transmitted through one leg. Therefore:

$$P_{RS} = p_a A_{ne}/\gamma_m$$
$$= 275 \text{ N/mm}^2 \times (100 \text{ mm} \times 6 \text{ mm})/1.2$$
$$= 137.5 \text{ kN}.$$

The limiting resistance is for flexural column buckling, giving:

$$P_R = 66.01 \text{ kN}.$$

7 Static design of members subjected to bending with axial force and to biaxial bending (4.8)

7.1 General (4.8.1)

For the following cases of combined actions, the interactions are as follows (4.8.1):

1) Case A, (M_x + P), major axis bending and an axial force,
2) Case B, (M_y + P), minor axis bending and an axial force,
3) Case C, (M_x + M_y), biaxial bending and
4) Case D, (M_x + M_y + P), biaxial bending and an axial force.

A section check is always needed (4.8.3). An overall buckling check (4.8.4) is required, but may not always be necessary for:

a) Case A if P is tensile and if the member is exempt from lateral torsional buckling (4.5.6.2) and
b) in Case B when P is tensile.

For the section check, account must be taken of any holes and any HAZ softening when determining P_{RS}, M_{RSx} and M_{RSy}.

P_{RS} is the factored axial tensile resistance of the section (4.6.2) or the factored axial compressive resistance of the section (4.7.7).

M_{RSx} is the factored uniaxial moment resistance of the cross-section, about a major axis, adjusted to allow for any coincident shear.

M_{RSy} is the factored uniaxial moment resistance of the cross-section, about a minor axis, adjusted to allow for coincident shear.

The values of M_{RSx} and M_{RSy} should refer to the most unfavourable section in the bay being considered. Account must be taken of HAZ softening and local buckling, but ignoring holes. HAZ softening occurring at the extreme ends of a spanning bay may be ignored (4.8.1.2).

M_{Rx} equals M_{RSx} for those sections defined in Sections 4.5.6.2 as being exempt from lateral torsional buckling.

M_{Rx} is the factored moment resistance to lateral torsional buckling (4.5.6.3).

7.2 Section classification and local buckling under combined actions (4.8.2)

For sections required to carry axial force with simultaneous bending or biaxial bending the section is given a single classification of fully compact, semi-compact or slender, in accordance with Sections 4.3.3.2, 4.3.3.4a or 4.8.2.1. When obtaining the axial resistance, if the section is found to be either fully compact or semi-compact it is considered compact and no reduction is made for local buckling (4.8.2.1).

For each element, the value of g required to modify β is found from Figure 4.2 and must correspond to the stress pattern produced when all the moments and the axial force are applied simultaneously. Figure 4.2 requires the quantities y_o and y_c. These quantities are usually found by using the elastic neutral axis of the gross section subjected to the combined actions. The plastic neutral axis may be used if a check is being made on whether a section is fully compact. Normally y_o and y_c will have different signs, but if the elastic neutral axis lies outside the section, they will have the same sign (4.8.2.1).

If the flange elements are under-stressed, the elements may be more favourably classified using Section 4.3.3.5, provided y_c and y_o relate to the stress pattern under the combined loading.

Where sections carrying biaxial bending or simultaneous axial force with bending have been classified as slender, each individual resistance must be found, using the relevant effective section for the action being considered (P, M_x or M_y). For example, k_L for each element under axial resistance relates to a value of β, corresponding to uniform stress ($g = 1$). The value of β for the moment resistance relates to the stress pattern in the element, due to the section being influenced by pure bending (4.8.2.2).

7.3 Section check (4.8.3)

The general formula that has to be satisfied for every position along the section length for biaxial bending and axial force of $M_x + M_y + P$, Case D of Section 4.8.1.1, is (4.8.3.1):

$$(P/P_{RS}) + (M_x/M_{RSx}) + (M_y/M_{RSy}) \leq 1.0$$

Each of the quantities in the equation is taken as positive. When considering Cases A, B and C of Section 4.8.1.1, the appropriate numerator quantities above are zero (4.8.3.2).

7.4 Overall buckling check (4.8.4)

When checking for overall buckling, any axial force is ignored when the member is subject to axial tension combined with bending. For axial compression with bending, or biaxial bending, the following formulae should be used for any unsupported length liable to buckle. All the quantities in the following equations are taken as positive (4.8.4.1).

For Case A, $(M_x + P)$, major axis bending and axial compression, the following two equations should be satisfied (4.8.4.2):

a) for the prevention of major axis buckling:

$$(P/P_{Rx}) + (\bar{M}_x/M_{RSx}) + (P\bar{M}_x/[2P_{Rx}M_{RSx}]) \leq 1.0$$

b) and for the prevention of minor axis buckling:

$$(P/P_{Ry}) + (\bar{M}_x/M_{Rx}) \leq 1.0$$

$\bar{M}_x$ is the equivalent uniform moment about the major axis as obtained in Section 4.5.6.4.

P_{Rx} is the factored axial resistance to overall column buckling about the major axis (4.7.3).

P_{Ry} is the factored axial resistance to overall column buckling about the minor axis (4.7.4).

For Case B, $(M_y + P)$, minor axis bending with axial compression, the following condition to prevent minor axis buckling must be satisfied (4.8.4.3):

$$(P/P_{Ry}) + (\bar{M}_y/M_{RSy}) + (P\bar{M}_y/[2P_{Ry}M_{RSy}]) \leq 1.0$$

$\bar{M}_y$ is the equivalent uniform moment about the minor axis (4.5.6.4).

For Case C, $(M_x + M_y)$, biaxial bending, to prevent minor axis buckling, the following condition must be satisfied (4.8.4.4):

$$(\bar{M}_x/M_{Rx}) + (\bar{M}_y/M_{RSy}) \leq 1.0$$

For Case D, $(M_x + M_y + P)$, biaxial bending and an axial force, the following condition must be satisfied:

$$(\bar{M}_x/M_{RPx}) + (\bar{M}_y/M_{RPy}) \leq 1.0$$

M_{RPx} is the lesser value of $\bar{M}_x$ that would be acceptable in combination with P, but in the absence of minor axis bending (4.8.4.2).

M_{RPy} is the value of $\bar{M}_y$ that would be acceptable in combination with P, but in the absence of major axis bending (4.8.4.4).

7.5 Design Example 1

An extruded symmetric I-section beam, 160 mm × 80 mm, in 6061-T6 material, has a span of 5 000 mm. The beam is free from HAZ softening and holes. It is known that: a) $V < V_{RS}$, b) the beam is fully compact and c) exempt from lateral torsional buckling. The beam is subjected to a combination of loads:

a) a factored dead load moment of 13.13 kN-m,
b) a factored live load of 5.83 kN and
c) an axial tension under factored loading of 200 kN.

Determine if the beam is satisfactory.

The beam is similar to that described in Chapter 6, at 6.9, 'Design example 1', but has an axial tensile force.

The beam is subjected to major axis bending plus an axial force $(M_x + P)$, (4.8.1.1a, Case A). The sum of the factored dead load maximum bending moment of 13.13 kN-m and the factored live load maximum bending moment of 5.83 kN-m is $M_x = 18.96$ kN-m. $P = 200$ kN, tension.

Two checks are needed, a section check (4.8.3) and an overall buckling check (4.8.4). The section check must always be carried out, but in this case where P is tensile and the member exempt from lateral torsional buckling, the overall lateral torsional buckling check may be waived (4.8.1.2).

For the section check, with all the quantities taken as positive:

$(P/P_{RS}) + (M_x/M_{RSx}) + (M_y/M_{RSy}) \leq 1.0$

There is no minor axis bending and $M_y = 0$. M_{RSY} need not be calculated (4.8.3.2).

$P_{RS} = p_o A/\gamma_m$ for general yielding along the member (4.6.2.2) or
$P_{RS} = p_a A_n/\gamma_m$ for local failure (4.6.2.3).

From Table 5 of BS 1161 for a 160 mm × 80 mm I-section, $A = 2\,820\ \text{mm}^2$ and $Z_x = 147\ \text{mm}^3 \times 10^3$. From Table 4.1 for 6061-T6, $p_o = 240\ \text{N/mm}^2$ and $p_a = 260\ \text{N/mm}^2$. The material factor $\gamma_m = 1.2$ (Table 3.3). $A_n = A$ as there are no holes in the section.

$P_{RS} = 240\ \text{N/mm}^2 \times 2\,820\ \text{mm}^2/1.2 = 564$ kN for general yielding or
$P_{RS} = 260\ \text{N/mm}^2 \times 2\,820\ \text{mm}^2/1.2 = 611$ kN for local failure,
giving $P_{RS} = 564$ kN.

As $V < 0.5V_{RS}$, M_{RS} is unaffected by coincident shear (4.5.4.1). In the absence of shear and for a fully compact section, $M_{RS} = p_o S_n/\gamma_m$ (4.5.2.2a) where:
S_n = the plastic modulus, which for an I-section $= 1.15Z_n$ and
$Z_n = Z_x$ as the section lacks holes,
giving $M_{RS} = 240\ \text{N/mm}^2 \times 1.15 \times 147\ \text{mm}^3 \times 10^3/1.2$
$= 33.81\ \text{kN-m} = M_{RSX}$ (4.8.1.2).

$(P/P_{RS}) + (M_x/M_{RSX}) + (M_y/M_{RSY})$
$= (200\ \text{kN}/564\ \text{kN}) + (18.96\ \text{kN-m}/33.81\ \text{kN-m}) + 0$
$= 0.915 \leq 1.0$

The section check is satisfied.

An overall buckling check is not required as the section is exempt from lateral torsional buckling and P is tensile [4.8.1.2].

8 Static design of members, deformation (4.9)

8.1 General (4.9.1)

The recoverable elastic deflection given in Section 4.9.2 due to nominal, unfactored loading should not exceed the limiting value specified in Section 3.4 (4.9.1). No separate check is needed for the in-service permanent inelastic deformation, which is assumed to be negligible, if the ultimate limit state static strength has been satisfied using Sections 4.2 to 4.8 inclusive (4.9.1).

8.2 Recoverable elastic deflections (4.9.2)

For compact sections and beams having either fully compact or semi-compact sections, the elastic deflection is calculated using the gross section properties and ignoring HAZ effects and holes (4.9.2.1).

For slender sections, the elastic deflection is calculated using the effective section properties and considering local buckling, but not HAZ softening or holes (4.9.2.2).

The assumed effective section, for slender sections, can be based on:

1) the reduced thickness as given in Section 4.5.2.3(1) for bending or
2) the reduced thickness as given in Section 4.7.6.3(1) for axial compression or
3) the more favourable of 4), 5) and 6).

4) The slender element may be reclassified, using a modified value for ϵ in Section 4.3.3.4, which is obtained by taking two-thirds of the normal value of p_0 given in Tables 4.1 and 4.2.
5) If 4) above reclassifies the section, then the gross section properties are used.
6) If 4) above still classifies the section as slender, another effective section is assumed for Figure 4.5, based on k_L determined using the modified value of ϵ found from 3) above.

9 An introduction to plates and plate girders (5.1)

9.1 General

A fabricated plate girder would normally be used to resist the high bending moments and high shearing forces that result from vertical loads supported over long spans. The flanges are required to resist the axial compressive and the axial tensile forces due to the bending actions of the vertical loads with the web resisting the shear forces. The connections between the flanges and the web transfer the longitudinal shear. Increasing the web depth reduces the axial forces in the flanges, but increases the likelihood of web buckling, unless the web is stiffened. Section 5 of BS 8118: Pt 1 covers the ultimate limit state of static strength for the following:

1) multi-stiffened plating as shown in Figure 5.2 of BS 8118: Pt 1 (5.3),
2) plate girders as shown in Figure 5.3 of BS 8118: Pt 1 (5.4) and
3) unstiffened plates as shown in Figure 5.1 of BS 8118: Pt 1 (5.2).

Section 4 of BS 8118: Pt 1 can be used to determine the resistance of plate girders and unstiffened plates. Because Section 4 uses simpler rules, the calculated resistance will not be as favourable as that obtained using Section 5 of BS 8118: Pt 1, especially for slender plates and for webs with low aspect ratios (a/d). Section 4 does not cover multi-stiffened plates (5.1).

9.2 Unstiffened plates under direct stress (5.2)

A plate subject to uniform in-plane compression P, as shown in Figure 9.1, is classified as:

1) fully compact if $\beta \leq \beta_1$,
2) semi-compact if $\beta_1 < \beta \leq \beta_o$ and
3) slender if $\beta > \beta_o$

where $\beta = (d/t)$ and β_o and β_1 are found from Table 4.3.
d is the depth of the unstiffened plate and t is the plate thickness (5.2.2.2).

For fully compact and semi-compact plates, P_{RS}, the factored axial resistance, is based on the least favourable cross-section (5.2.2.3):

1) for fully compact sections $P_{RS} = (p_a A_{ne}/\gamma_m)$ and
2) for semi-compact sections $P_{RS} = (p_o A_{ne}/\gamma_m)$.

p_a and p_o are the limiting stresses found from Tables 4.1 and 4.2. A_{ne} is the net effective area after allowing for holes and taking into consideration a reduced thickness $k_z t$, for any region affected by HAZ softening (4.4.2 and 4.4.3).

For slender plates, a check on buckling and yielding is required, with P_{RS}, the factored axial resistance, being found as follows (5.2.2.4):

1) for the buckling check, $P_{RS} = (p_o A_e/\gamma_m)$ and
2) for the yielding check, $P_{RS} = (p_o A_{ne}/\gamma_m)$, but ignoring buckling.

p_o is the limiting stress found from Tables 4.1 and 4.2, with A_e being the effective area determined by using the reduced thickness to allow for buckling and to allow for HAZ softening, but ignoring holes (5.2.2.4).

In the buckling check, the effective area is based on the least favourable cross-section, using a thickness of $k_z t$ or $k_L t$ in the HAZ region, whichever is the lower, and $k_L t$ everywhere else. If weld HAZ softening occurs at the loaded edges, it may be ignored.

The local buckling coefficient k_L is the most favourable of either:

1) k_L is found from Figure 4.5, curves C or D, taking $\beta = d/t$ and $\epsilon = (250/p_o)^{0.5}$, if the plate treatment is used, or
2) $k_L = p_s/p_o$ if the column treatment is used. p_s is the column buckling stress read from Figure 4.10a. The correct curve on Figure 4.10a is that which intersects the stress axis at a value p_o. Usually $\lambda = (3.5a/t)$, corresponding to a simple support. A lower value may be used if it is realistic (5.2.2.4).

9.2.1 Unstiffened plates under in-plane moment (5.2.3)

A plate subject to a pure in-plane moment as shown in Figure 9.1a, acting on the sides of width d, is classified as:

1) fully compact if $\beta \leq \beta_1$,
2) semi-compact if $\beta_1 < \beta \leq \beta_o$ and
3) slender if $\beta > \beta_o$ where

$\beta = (0.35d/t)$ and β_o and β_1 are found from Table 4.3 (5.2.3.2).

If the moment varies in the direction parallel to dimension a, then Section 5.2.4 is used.

For fully compact or semi-compact plates, M_{RS}, the factored moment resistance in the absence of shear, is based on the least favourable cross-section, using the appropriate expression for M_{RS} found from Section 4.5.2.2a to d. The assumed section is the net section or the net effective section as defined in Section 4.5.2.3a or b (5.2.3.3).

For slender plates, M_{RS} is the lower of the following two values:

1) $M_{RS} = (p_o Z_e/\gamma_m)$, for the buckling check, where Z_e is the elastic modulus of the effective section, or
2) M_{RS}, for the yielding check, is found as described above for a semi-compact plate, but ignoring buckling, and is based on the least favourable cross-section, using the appropriate expression for M_{RS} found from Section 4.5.2.2a to d. The assumed section is the net section or the net effective section as defined in Section 4.5.2.3a or b (5.2.3.3).

In the buckling check, the effective section is found at the least favourable position, with no deduction for holes, using a thickness of $k_z t$ or $k_L t$ in the HAZ region, whichever is the lower, and $k_L t$ everywhere, else. If weld HAZ softening occurs at the loaded edges, it may be ignored. k_L is read from Figure 4.5, curves C or D, using $\beta = (0.35d/t)$ and $\epsilon = (250/p_o)^{0.5}$.

9.2.2 Longitudinal stress gradient on unstiffened plates (5.2.4)

This section is used when either M or P in an unstiffened plate varies longitudinally, that is, in the direction parallel to the dimension a in Figure 9.1. For fully compact or semi-compact plates, the factored resistance must be equal to or more than the action at that section due to factored loading. For slender plates, every cross-section must satisfy the yielding check. However, for the buckling check, it is only necessary to compare the factored resistance with the action that arises at 0.4 times the elastic plate buckling half-wavelength from the more heavily loaded end of the panel (5.2.4.3).

9.2.3 Unstiffened plates in shear (5.2.5)

Unstiffened plates in shear are classified as compact or slender, as described in Section 4.5.3.1. Provided any small holes do not occupy more than 20 per cent of the cross-sectional area across the width d, the holes may be ignored when finding the shear resistance (5.2.5.1).

For a compact shear web, V_{RS}, the factored shear force resistance, is as follows (5.2.5.2):

$V_{RS} = (p_v A_v / \gamma_m)$ where

A_v is the effective shear area and p_v is the limiting stress in shear found from Tables 4.1 and 4.2. A_v is found as follows:

1) for partially welded plates, A_v is the effective area on the width d, determined by taking the reduced thickness $k_z t$ in the softened zones (4.4.2 and 4.4.3),
2) for plates fully welded along one or more edges, $A_v = k_z dt$ and
3) for unwelded plates, $A_v = dt$.

For a slender shear web, V_{RS} is the lesser of the following two values (5.2.5.3):

1) for the yielding check
   $V_{RS} = (p_v A_v / \gamma_m)$ as described for the compact plate above (5.2.5.2) or
2) for the buckling check
   $V_{RS} = ([340\ Nt^3]/[d\gamma_m])$ as in Section 4.5.3.3b or
   $V_{RS} = ([v_1 p_v dt]/[\gamma_m])$ which is more favourable if $a < 2.5d$.

v_1 is the elastic critical shear buckling factor found from Figure 5.4, using $\epsilon = (150/p_v)^{0.5}$.

The use of $V_{RS} = ([340\ Nt^3]/[d\gamma_m])$ as in Section 4.5.3.3b does not consider the tension field action advantage. If such an advantage is sustainable due to the edge conditions, then Section 5.4.3.4 for type 1 panels in plate girder webs may be used.

9.2.4 Combined actions (5.2.6)

When a plate is subjected to both an axial force P and a moment M under factored loading, the plate must be given a single classification of fully compact, semi-compact or slender in accordance with Section 4.8.2.1. The value of β is based on the combined stress pattern in the plate due to P and M acting together and on the appropriate value of g in Figure 4.2. If the plate is slender, then the individual resistances P_{RS} and M_{RS} are based on the specific type of action being considered as in Section 4.8.2.2 (5.2.6.1).

For a plate subjected to an axial force and a moment (5.2.6.2):

$$(P/P_{RS}) + (M/M_{RS}) \leq 1.0$$

where:
P is the axial force due to factored loading,
M is the in-plane moment due to the factored loading,
P_{RS} is the factored resistance to axial force reduced to allow for any coincident high shear (5.2.6.4) and
M_{RS} is the factored resistance to in-plane moment reduced to allow for any coincident high shear (5.2.6.4).

If $V \leq 0.5V_{RS}$, then V, the coincident shear force under factored loading, has no effect on the longitudinal resistance of the plate (5.2.6.3). If $V > 0.5V_{RS}$, the longitudinal resistance of axial force or moment must be reduced by k_v, where (5.2.6.4):

$$k_v = 1.6 - 1.2(V/V_{RS}).$$

9.3 Multi-stiffened plating (5.3)

A typical multi-stiffened plate is shown in Figure 5.2 of BS 8118: Pt 1. It is to be supported on all four edges, that is, reinforced with three or more equally spaced corrugations or longitudinal stiffeners. The corrugations or longitudinal stiffeners can be unsupported on their whole length or be continuous over intermediate transverse stiffeners. The spacing between supports, when fitted, is L. An essential feature of the design of multi-stiffened plates is that the longitudinal reinforcement is sub-critical and can deform with the plating in an overall buckling mode.

Sections 5.3.2, 5.3.3 and 5.3.4 deal with the resistance of multi-stiffened plating to longitudinal direct stress in the direction of the reinforcement. Section 5.3.5 considers shear. Section 5.2.6 is used to consider interaction between the different effects. If the cross-section contains any outstand elements that are classified as slender, then these sections cannot be used.

Where the multi-stiffened plating is flat plating with applied stiffeners, then the resistance to transverse direct stress is the same as for an unstiffened plate. Corrugated multi-stiffened plating has negligible resistance to transverse direct stress (5.3.1).

9.3.1 Multi-stiffened plating under uniform compression (5.3.2)

The cross-section is classified as compact or slender (4.3.3), having considered all the component elements. Slender outstand elements are not allowed. A column check (5.3.2.3) and a yielding check (5.3.2.2) need to be made (5.3.2.1).

For the yielding check, the entire section should be checked for local squashing as for a strut (4.7.7). P_{RS} is based on the least favourable cross-section, taking into account any HAZ softening, local buckling and unfilled holes (5.3.2.2).

For the column check, the plating is regarded as an assemblage of identical column sub-units, each having one centrally located stiffener or corrugation with a pitch w equal to the width between the stiffeners. P_{RS} is as follows:

$$P_{RS} = (p_s A/\gamma_m)$$

where A is the gross area of the entire cross-section of the plating. p_s is read from Figure 4.10 and is the appropriate buckling stress for a column sub-unit, treated as the column buckling of a simple strut out of plane of the plating. λ for Fig 4.10 is based on the lesser of the two effective lengths l, found as follows:

1) the distance between the positions of effective lateral support, e.g. the effective transverse stiffeners or the end supports or
2) the elastic orthotropic buckling half-wavelength.

The section shape of the sub-unit, and whether it contains longitudinal welding as described in Table 4.10, affects whether Figure 4.10a, b or c is used. The correct curve is that which intercepts the stress axis at a value of p_l as defined in Section 4.7.6.2.

When determining A_e, k_L for elements such as E in Figure 5.2 is based on their full dimensions, even if they are cut in two for the formation of sub-units. Further, HAZ softening due to welds at the loaded edges or at transverse stiffeners may be ignored (5.3.3.2).

9.3.2 Multi-stiffened plating under in-plane moment (5.3.3)

For multi-stiffened plating under in-plane moment it is necessary to classify the cross-section as either compact or slender, before carrying out a column check and a yielding check (5.3.3.1). When classifying individual elements and when determining k_L for slender elements, it is assumed that each element is under uniform compression and that $g = 1$ in Section 4.3.2.2. For the yielding check, g is based on the true stress pattern in the elements making up the outermost region of the plating. This value is used for corresponding elements further in and can be favourable when there are few corrugations or stiffeners. Slender outstand elements are not allowed (5.3.3.2).

For the yielding check, the entire cross-section of the plating is treated as a beam under in-plane bending (4.5.2.2), with M_{RS} based on the least favourable cross-section and considering HAZ softening, holes and local buckling (5.3.3.3).

For the column check, the plating is regarded as an assemblage of column sub-units, each having one centrally located stiffener or corrugation with a pitch w equal to the width between the stiffeners, as for axial compression (5.3.2.3), with

$$M_{RS} = ([p_s ZB]/[2y\gamma_m])$$

where Z is the elastic section modulus of the full cross-section of the plating for in-plane bending and p_s being read from Figure 4.10 in the same way as for uniform compression in Section 5.3.2.3 (5.3.3.4).

9.3.3 Longitudinal stress gradient on multi-stiffened plates (5.3.4)

Where M or P vary in the direction of the corrugations or the stiffeners on a multi-stiffened plate, a column check and a yielding check are required (5.3.4). For the column check, it is only necessary to compare the factored resistance with the action due to factored loading at a distance $x = 0.4$ times the effective buckling length l, from the more heavily loaded end of a panel (5.3.4.3). For the yielding check, the factored resistance at any cross-section should be equal to or greater than the action arising at that section under factored loading (5.3.4.2).

9.3.4 Multi-stiffened plating in shear (5.3.5)

For multi-stiffened plating in shear, the buckling and yielding checks are valid only if the following apply (5.3.5.1):

1) $w \leq 0.3L$,
2) outstand elements of the section are classified as compact in terms of axial resistance (4.3.3.4),
3) internal elements are classified as compact in terms of the shear resistance (4.5.3.1) and

4) all corrugations or stiffeners and the actual plating are continuous at any transverse stiffener position and are effectively connected to the transverse framing at either end.

For the buckling check, no allowance is needed for HAZ softening and

$V_{RS} = (v_1 p_v B t)/(\gamma_m)$.

However, it is necessary to calculate v_1, the elastic critical shear buckling factor as found in Figure 5.4, using the following:

a $= B$,
d $= 0.61([wt^3]/[I_{SU}])^{0.375}$,
l = the effective length of the plating and is taken as the unsupported length L in Figure 5.2,
w = the pitch between the corrugations or the stiffeners,
I_{SU} = the second moment of area of one sub-unit of the plating as defined in Section 5.3.2.3, about a centroidal axis parallel to the plane of the plating and
$\epsilon = (150/p_v)^{0.5}$.

When $L \gg B$, it may be more favourable to let l be equal to the elastic orthotropic shear buckling half-wavelength (5.3.5.3).

For the yielding check, V_{RS} is assumed to be the same as for a flat unstiffened plate of the same general aspect of $L \times B$ and with the same general thickness t found using Section 5.2.5.2 (5.3.5.2).

9.4 Design example 1

An unstiffened, welded, internal plate element, of width a = 1000 mm, depth d = 250 mm and thickness t = 10 mm, is subjected to a uniform axial compressive force due to factored loading of P = 25 kN (Figure 9.1 overleaf). P is applied to the 250 mm long edges through plates welded to those edges. All the plate material is 6082-T4. Determine if the plate has the required factored resistance P_{RS}.
Table 3.3 gives:

$\gamma_m = 1.2$, while Table 4.1 for 6082-T4 plate material gives:
$p_o = 105$ N/mm^2,
$p_a = 140$ N/mm^2 and
$p_v = 65$ N/mm^2.

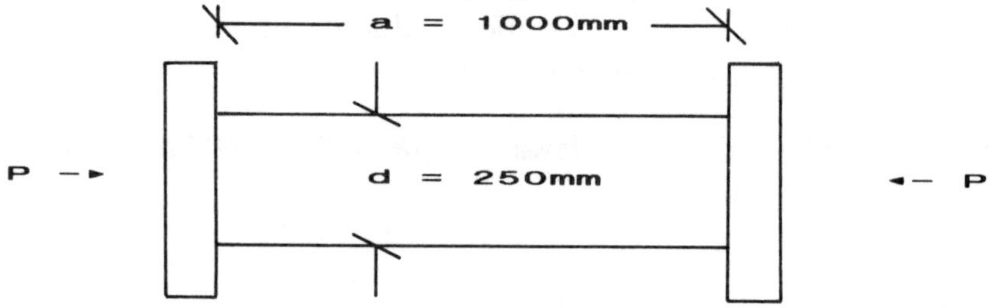

Figure 9.1 Unstiffened plate subject to a uniform compression

Although Section 4 of BS 8118: Pt 1 could be used to obtain the resistance of the plate, Section 5 may give more favourable results (5.1).

The plate must be classified as:

fully compact $\beta \leq \beta_1$,
semi-compact $\beta_1 < \beta \leq \beta_o$ or
slender if $\beta > \beta_o$ (5.2.2.2).
$\beta = d/t = 250\ mm/10\ mm = 25$.

Using Table 4.3 and Figure 4.1 for welded internal elements,

$\beta_o = 18\epsilon$ and
$\beta_1 = 15\epsilon$ where
$\epsilon = (250/105\ N/mm^2)^{0.5}$ giving
$\beta_o = 27.77$ and
$\beta_1 = 23.15$, giving
$\beta_1\ (23.15) < \beta\ (25) \leq \beta_o\ (27.77)$.

The plate is semi-compact.

For semi-compact plates:

$P_{RS} = p_o A_{ne}/\gamma_m$ (5.2.2.3) where
$A_{ne} = k_z \times 250\ mm \times 10\ mm$ to take account of the HAZ softening about the weld (5.2.2.3).
k_z is found from Table 4.5, which for 6082-T4 is 1.0 (4.4.2) giving
$P_{RS} = 105\ N/mm^2 \times 1.0 \times 250\ mm \times 10\ mm/1.2$
$= 218.75$ kN.

The plate has the required factored resistance as P_{RS} (218.75 kN) > P (25 kN).

9.5 Design example 2

An unstiffened, welded, internal plate element, of width a = 1 000 mm, depth d = 250 mm and thickness t = 10 mm, is subjected to a pure in-plane moment due to factored loading of M = 10 kN-m (Figure 9.2 below). M is applied to the 250 mm long edges through plates welded to those edges. All the plate material is 6082-T4 plate. Determine if the plate has the required factored resistance M_{RS}.

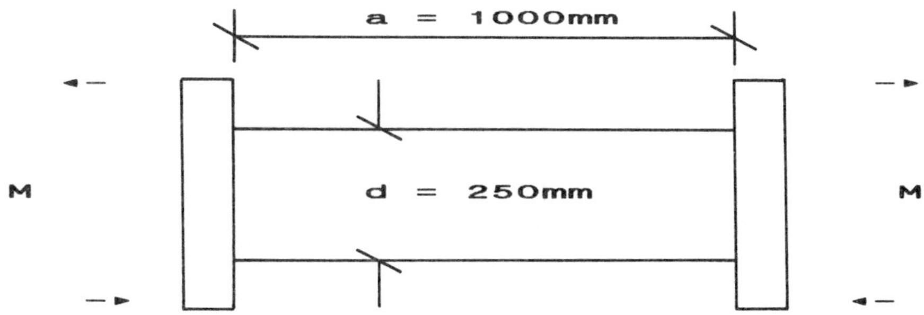

Figure 9.2 Unstiffened plate subject to pure in-plane moments

Table 3.3 gives:

$\gamma_m = 1.2$.
Table 4.1 for 6082-T4 plate gives
$p_o = 105$ N/mm^2,
$p_a = 140$ N/mm^2 and
$p_v = 65$ N/mm^2.

Section 4 of BS 8118: Pt 1 could be used to obtain the resistance of the plate, but Section 5 may give more favourable results (5.1).

The plate must be classified as:

fully compact $\beta \leq \beta_1$,
semi-compact $\beta_1 < \beta \leq \beta_o$ or
slender if $\beta > \beta_o$ (5.2.2.2).
$\beta = 0.35d/t$
$= 0.35 \times 250$ mm/10 mm
$= 8.75$ (5.2.3.2).

Using Table 4.3 and Figure 4.1 for welded internal elements,

$\beta_o = 18\epsilon$ and
$\beta_1 = 15\epsilon$ where
$\epsilon = (250/105\ \text{N/mm}^2)^{0.5}$ giving
$\beta_o = 27.77$ and
$\beta_1 = 23.15$ which gives
$\beta\ (8.75) < \beta_1\ (23.15)$.

The plate is fully compact.
M_{RS} is based on the least favourable cross-section (5.2.3.3). For a welded fully compact section:

$M_{RS} = p_oS_{ne}/1.2$
$= 105\ \text{N/mm}^2 \times S_{ne}/1.2$ (4.5.2.2c).
S_{ne} is the plastic modulus, which for a rectangular section is $1.5 \times Z_{ne}$.
$Z_{ne} = 1.5 \times t \times d^2/6$, but $t = k_zt$, due to HAZ effects (4.5.2.3).
$k_z = 1.0$ (Table 4.5 and 4.4.2.1) giving
$S_{ne} = 1.5 \times 1.0 \times 10\ \text{mm} \times (250\ \text{mm})^2/6$
$= 156.25\ \text{mm}^3 \times 10^2$ and
$M_{RS} = 13.67$ kN-m.

As M_{RS} (13.67 kN-m) > M (10 kN-m), the plate has the required factored resistance M_{RS}.

If M in this design example, or P in 9.4 'Design example 1', varies longitudinally along the plate width, that is, in the a = 1 000 mm direction, M_{RS} or P_{RS} respectively at any section, for fully compact and semi-compact plates, must be greater than M or P respectively at that section (5.2.4.2). For slender plates, the yielding check must be satisfied at every cross-section. The buckling check is made at 0.4 times the elastic plate buckling half-wavelength from the more heavily loaded end of the panel.

9.6 Design example 3

An unstiffened, welded, internal plate element, of width a = 1 000 mm, depth d = 250 mm (750 mm in the second part of this design example) and thickness t = 10 mm, is subjected to a shear force under factored loading of V = 40 kN (Figure 9.3 opposite). V is applied to the 250 mm long edges through plates welded to those edges. The plate material is 6082-T4. Determine if the plate has the required factored shear force resistance.

Table 3.3 gives $\gamma_m = 1.2$ and Table 4.1 gives $p_o = 105\ \text{N/mm}^2$, $p_a = 140\ \text{N/mm}^2$ and $p_v = 65\ \text{N/mm}^2$.

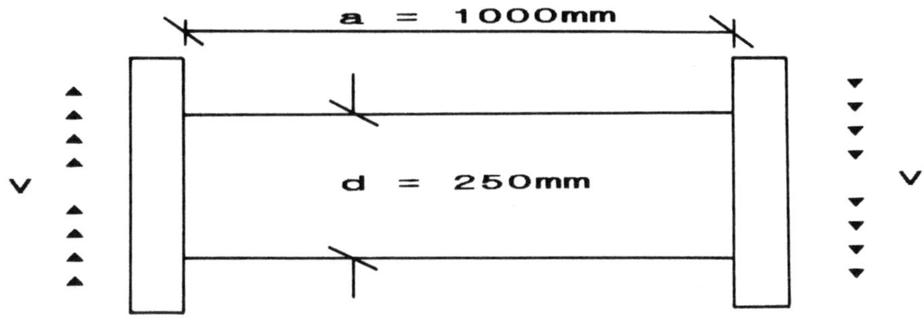

Figure 9.3 Unstiffened plate under shear loading

The plate must be classified as compact or slender (5.2.5.1). The section is compact if

$d/t \leq 49\epsilon$ and slender if
$d/t > 49\epsilon$ (4.5.3.1(1)).
$d/t = 250$ mm/10 mm = 25.
$\epsilon = (150/65 \text{ N/mm}^2)^{0.5}$
$= 1.519$ (4.5.3.1), giving
$49\epsilon = 74.44$.
d/t (25) $< 49\epsilon$ (74.44), therefore the section is compact.

For the compact shear web:

$V_{RS} = p_v A_v/\gamma_m$
$= 65 \text{ N/mm}^2 A_v/1.2$, where
$A_v = k_z \times 250 \text{ mm} \times 10 \text{ mm}$ and
$k_z = 1.0$ (4.4.2.1 and Table 4.5) giving
$V_{RS} = 135.42$ kN.

As V_{RS} (135.42 kN) $>$ V (40 kN), the plate is satisfactory and has the required factored shear resistance.

If the plate had been slender, that is, $d/t > 74.44$, say 75 giving d = 750 mm, then V_{RS} would be the lesser of the values found from a buckling check and a yielding check (5.2.5.3).

The buckling check uses Section 4.5.3.3b (5.2.5.3).

$V_{RS} = 340 \text{ N}t^3/(d\gamma_m)$. d is the clear depth of the web, between the flanges.
$= (340 \times 1 \times [10 \text{ mm}]^3)/(750 \text{ mm} \times 1.2)$
$= 377.78$ kN.

Further, the following equation may give a more favourable result when $a < 2.5d$, (1 000 mm) < (2.5 × 750 mm)

$V_{RS} = v_l p_v dt/\gamma_m$
$= v_l \times 65\ \text{N/mm}^2 \times 750\ \text{mm} \times 10\ \text{mm}/1.2$

v_l is found from Figure 5.4 where
$\epsilon = (150/65\ \text{N/mm}^2)^{0.5}$
$= 1.519$
$d/(t\epsilon) = 750\ \text{mm}/(10\ \text{mm} \times 1.519)$
$= 49.37.$
$a/d = 1\,000\ \text{mm}/750\ \text{mm}$
$= 1.33$, giving
$v_l = 1.0$ and
$V_{RS} = 406.25$ kN.

Using Section 4.5.3.3b gives $V_{RS} = 377.78$ kN which is a conservative value as suggested in Section 5.2.5.3. When $a < 2.5d$ as in this example, the equation for V_{RS} given in Section 5.2.5.3b gives $V_{RS} = 406.25$ kN; a more favourable result.

For the yielding check:

$V_{RS} = p_v A_v/\gamma_m$
$= 65\ \text{N/mm}^2 A_v/1.2$, where
$A_v = k_z \times 750\ \text{mm} \times 10\ \text{mm}$ and
$k_z = 1.0$ (4.4.2.3 and Table 4.5) giving
$V_{RS} = 406.25$ kN.

The value of V_{RS} is 406.25 kN, in both buckling and yielding. V_{RS} (406.25 kN) > V (40 kN), thus the section has the required factored shear force resistance.

9.7 Design example 4

The web element of an unwelded built-up girder consists of a multi-stiffened plate supported on all four edges and reinforced with at least three equally spaced longitudinal corrugations as shown in Figure 9.4 opposite. The web has an overall width of B = 900 mm and a length between supports of L = 400 mm. The pitch of the stiffeners is w = 150 mm, and the corrugation depth is 50 mm. The plate thickness is 2.65 mm, the total cross-sectional area is 3 074 mm^2, that is, 535.3 mm^2 per corrugation. I = 148 028 mm^4 per 150 mm sub-unit

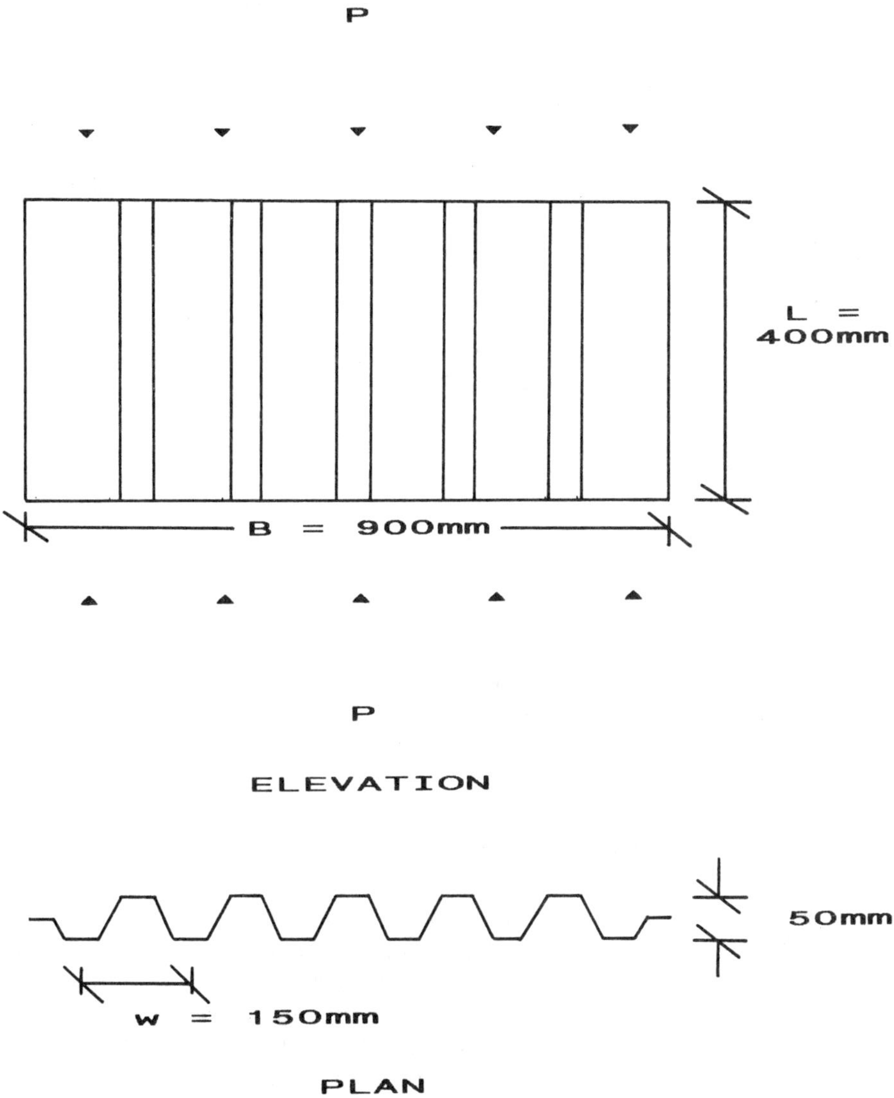

Figure 9.4 Multi-stiffened plate under compressive load

(corrugation). The alloy is 3103-H18, with $p_o = p_a = 150$ N/mm^2 and $p_v = 90$ N/mm^2 (Table 4.2). $\gamma_m = 1.2$ (Table 3.3). Determine P_{RS} if the web is considered unwelded with no holes.

As the plate is of corrugated construction, the resistance to transverse direct stress is negligible (5.3.1). The cross-section contains no outstand elements that could be classified as slender. Each end of the overall width B = 900 mm, is supported by two equal angle vertical post sections (5.3.1).

For uniform compression a column check (5.3.2.3) and a yielding check (5.3.2.2) are required (5.3.2.1).

The section must be classified as compact or slender to Section 4.3.3 (5.3.2.3).

Each element (sub-unit) is considered flat, internal and unreinforced. For axial resistance:

$\beta \leq \beta_o$ for a compact section and
$\beta > \beta_o$ for a slender section (4.3.3.4).

For unwelded internal elements:

$\beta_o = 22\epsilon$, where
$\epsilon = (250/p_o)^{0.5}$ giving
$\beta_o = 22(250/150\ N/mm^2)^{0.5}$
$= 28.40$ (Table 4.3).

Under uniform compression $\beta = b/t$ (4.3.2.2). b has two values, 61 mm and 50 mm, giving a maximum value of

b = 61 mm (Figure 9.5), giving
$\beta = 61\ mm/2.65\ mm$
$= 23.02 \leq \beta_o$ (28.40).

The element is compact and local buckling can be ignored (4.3.3.2).

For the column check each 150 mm corrugation sub-unit is considered, but A is for the entire cross-section of the plating and $P_{RS} = p_s A/\gamma_m$ (5.3.2.3). For the sub-unit, the pitch w = 150 mm and A per 150 mm corrugation = (61 mm + 40 mm + 61 mm + 40 mm) × 2.65 mm = 535.3 mm. p_s is found from Figure 4.10 using Section 4.7.6. The sub-unit is unwelded and mildly asymmetric, thus Figure 4.10a is required (Table 4.10).

For compact sections, with no HAZ effects, $p_l = p_o = 150\ N/mm^2$ (4.7.6.2). $\lambda = l/r$ (4.7.4.2). l = L = 400 mm, which we can reasonably assume to be less than the elastic orthotropic buckling half-wavelength in the L direction (5.3.2.3).

To determine r per sub-unit $I = Ar^2$, where $I = 148\ 028\ mm^4$, giving

$r = (I/A)^{0.5}$
$= (148\ 028\ mm^4/535.3\ mm^2)^{0.5}$
$= 16.629$ mm giving
$\lambda = l/r$
$= 400$ mm/16.888 mm
$= 24.05.$

From Figure 4.10a, using $p_l = p_o = 150\ N/mm^2$, going up the y axis and interpolating between the 140 N/mm^2 and the 160 N/mm^2 lines, until one crosses the $\lambda = 24.05$ line, giving $p_s = 143.1\ N/mm^2$ for the column sub-unit. Therefore

$P_{RS} = 143.1\ N/mm^2 \times 3\ 074\ mm^2/1.2$
$= 366.46$ kN.

For the yielding check (5.3.2.2), there is no HAZ effect, no unfilled holes and local buckling can be ignored. In this case $A_n = A$.

$P_{RS} = p_a A_n/\gamma_m$
$= 150\ N/mm^2 \times 3\ 074 mm^2/1.2$
$= 384.25$ Kn if spread along the entire 900 mm overall width B (4.7.7).

The lowest value of P_{RS} for the entire cross-section of the plating is 366.46 kN for the column check.

Normally the value of the elastic orthotropic buckling half-wavelength in the L direction is calculated using a standard text such as *Theory of Elastic Stability* by Timoshenko and Gere (McGraw-Hill, New York, 2nd ed., 1961, Chap 9.11) or perhaps obtained from the manufacturer's data sheets. Essentially the half-wavelength is related to the edge restraints, the L/B ratio and to the flexural rigidity of the stiffeners. A long plate may buckle into several half-waves, with a single half-wave usually giving the lowest critical loading. The length of the half-wave increases as the flexural rigidity of the stiffener increases. There are also half-wavelengths of a different magnitude across the width of the plate, but BS 8118 does not refer to these.

9.8 Design example 5

The web element of an unwelded built-up girder consists of a multi-stiffened plate supported on all four edges and reinforced with equally spaced longitudinal corrugations. The web has a total horizontal width of a = B = d = 900 mm and a length (height) between supports (flanges)

of $l = L = 510$ mm. The pitch of the stiffeners is $w = 150$ mm and the corrugation depth is 50 mm (Figure 9.5 opposite). The corrugated plate thickness is 8.1 mm. $I_{SU} = 452\,463$ mm^4 per corrugation. The alloy is 3103-H18, with $p_v = 90$ N/mm^2 (Table 4.2). $\gamma_m = 1.2$ (Table 3.3). The corrugations are effectively connected to the transverse framing at either end, the flanges (5.3.5.1d(1)). Determine V_{RS} of the web if a shear force is applied to the two horizontal flanges.

For multi-stiffened plating under shear load a buckling check and a yielding check are required. The following must also be satisfied (5.3.5.1):

1) The pitch w (150 mm) $\leq$ 0.3L (0.3 $\times$ 510 mm = 153 mm), therefore satisfactory (5.3.5.1a),
2) the two outstand elements at the ends of the overall width are supported by two equal angle vertical end-post sections and are thus classified (as required) as compact regarding axial resistance (5.3.5.1b) and
3) for shear resistance, the vertical internal elements must be classified as compact and comply with Section 4.5.3.1 (5.3.5.1c).
   Therefore d/t (510 mm/8.1 mm = 62.96) $\leq 49\epsilon$ ($49 \times (150/90$ N/mm$^2)^{0.5}$ = 63.26).
   The section is thus compact and is satisfactory (5.3.5.1c).

For the buckling check:

$V_{RS} = v_1 p_v B t/\gamma_m$
$= v_1 \times 90$ N/mm$^2 \times 900$ mm $\times$ 8.1 mm/1.2, with v_1 being found from Figure 5.4.
$d = 0.6 \times 510$ mm $\times$ ([150 mm $\times$ {8.1 mm}^{3}]/452 463 mm^4)$^{0.375}$
$= 159.57$ mm. $a = B = 900$ mm. For Figure 5.4:
$(d/[t\epsilon]) = 159.57$ mm/(8.1 mm $\times$ 1.291)
$= 15.26$ and
$a/d = 900$ mm/159.57 mm
$= 5.64$ giving
$v_1 = 1$ and
$V_{RS} = 546.75$ kN.

For the yielding check Section 5.2.5.2 is used (5.3.5.2).

$V_{RS} = p_v A_v/\gamma_m$, but according to Section 5.2.5.2, $A_v = dt$; however, for the unstiffened unwelded plate, d is redefined as the plate width and is equal to B for multi-stiffened plates, giving $A_v = dt = Bt = 900$ mm $\times$ 8.1 mm. Therefore, $V_{RS} = p_v A_v/\gamma_m$

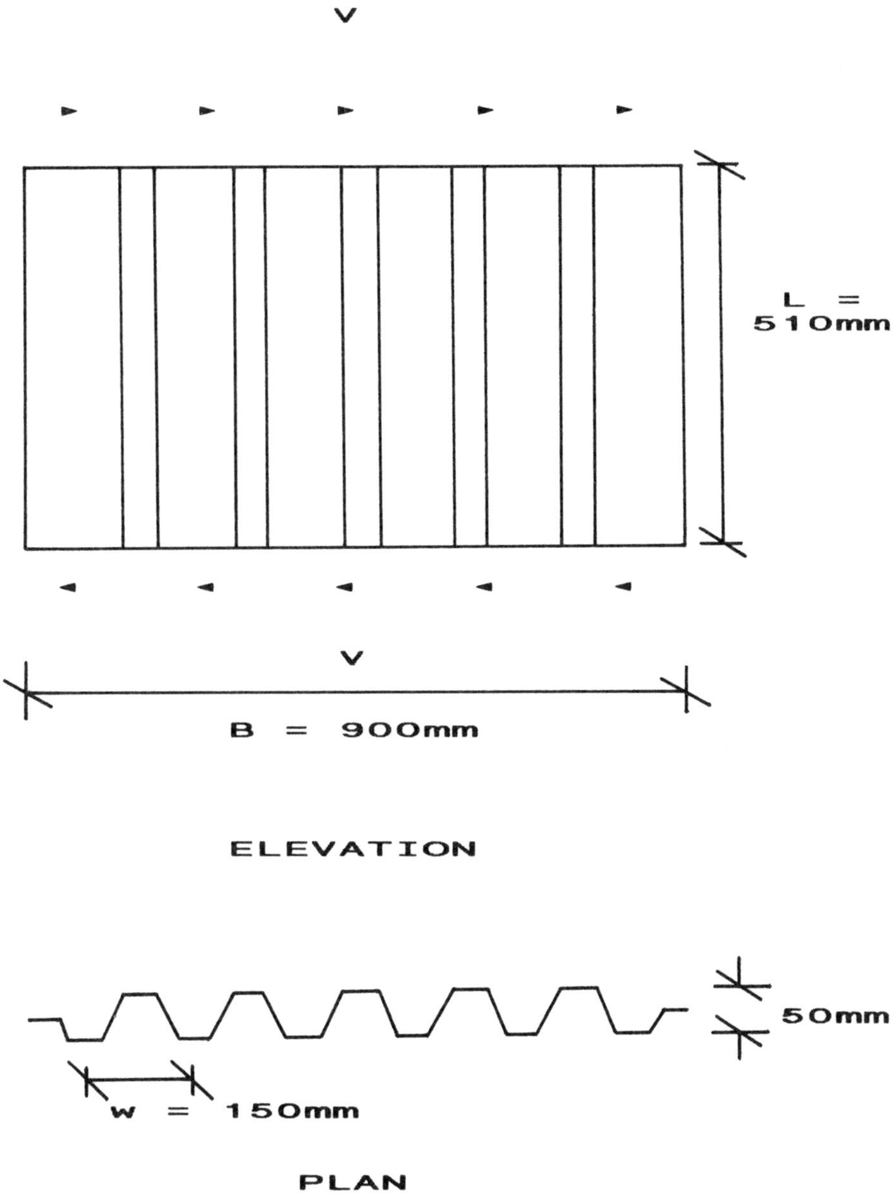

Figure 9.5 Multi-stiffened plate under shear load

$$= 90 \text{ N/mm}^2 \times (900 \text{ mm} \times 8.1 \text{ mm})/1.2$$
$$= 546.75 \text{ kN.}$$

Care must be taken in this example over the use of different letters used to represent the same quantity; e.g. d = a = B. Also, the same letter, in this case d, is used to define two different quantities.

10 Plate girders (5.4)

10.1 General

A plate girder is a fabricated beam having a compression flange, a tension flange and a web plate. The web is usually of slender proportions, reinforced transversely with bearing and intermediate stiffeners as shown in Figure 5.3 of BS 8118: Pt 1. Longitudinal stiffeners may also be incorporated. The web stiffeners provide supported edges for the panels of the web and stay essentially straight as buckling proceeds. There may be advantages in providing a tongue plate to one or both flanges. Tongue plates should comply to Section 5.4.5.8.

The calculations to determine the moment and the shear resistance of plate girders with transversely stiffened webs and with added longitudinal stiffeners are valid providing the following are satisfied (5.4.1):

1) the stiffeners comply with Section 5.4.5 and
2) a, the transverse stiffener spacing, is equal to or greater than half of the clear depth of the web between the flange plates. For corrugated or closely stiffened webs Section 5.4.6 should be consulted.

Lateral torsional buckling or web bearing may need to be considered (4.5.5 and 4.5.6); further, for girders subjected to axial load and bending, Section 4.8 should be consulted. Also, this chapter for plate girders is, in general, applicable to box section girders with similar forms of webs.

10.2 Moment resistance of transversely stiffened plate girders (5.4.2)

It is necessary to carry out a buckling check and a yielding check to determine the factored moment resistance. If the girder is hybrid, in that the flange and web materials are different, Section 4.5.2.4 should be consulted.

For the yielding check, the moment at any section due to factored loading should be equal to or less than the value of M_{RS}, found if the section were treated as semi-compact. M_{RS} is found using Section 4.5.2.2b or d, taking into account HAZ effects and holes but ignoring local buckling (5.4.2.2).

For plate girders with transverse stiffeners, but not with longitudinal stiffeners, the following buckling check is made (5.4.2.3). The moment due to factored loading that arises in each bay, between transverse stiffeners, at a distance of 0.4a from the more heavily stressed end, should not exceed M_{RS} for that bay, where M_{RS} is based on the ultimate failure by buckling. M_{RS} is found from Section 4.5.2.2e, allowing for HAZ softening and local buckling, but ignoring holes. HAZ effects due to the welding of transverse stiffeners can be ignored.

For web buckling k_L, the effective thickness factor, is found using Section 4.3.4. β is found from Section 4.3.2.2. If the tension edge of the web is further from the neutral axis than the compression edge, then Section 5.4.2.4 can be used and may be more favourable. Also, a tongue plate provides effective edge support to a slender web plate to which it is joined if it complies with Section 5.4.5.8, and thus β is based on d measured to the tip of the tongue or tongues (5.4.2.3).

If the compression edge of the web is nearer to the neutral axis than the edge in tension, then the web is treated as two zones with two values of k_L found as follows (5.4.2.4):

1) for zone 1, which extends y_1 either side of the neutral axis, k_L is found from Figure 4.5b, with $\beta = (0.7y_1/t)$, with y_1 equal to the distance from the gross neutral axis to the compression edge and with
2) $k_L = 1.0$ for zone 2, the rest of the web.

When considering the moment resistance the elasto-plastic moment calculations of Appendix E may be used rather than Section 5.4.2. The elasto-plastic method has advantages for sections where a critical element for local buckling has a value of β similar to β_o and also has the outermost tension fibres further from the neutral axis than the compression fibres; for example, where a critical web element stops some distance from the extreme compression face, due to a tongue plate.

10.3 Shear resistance of transversely stiffened plate girders (5.4.3)

To determine the shear resistance of transversely stiffened plate girders, a buckling check and a yielding check is required. A HAZ check is also required for a web with a continuous longitudinal weld. Provided any small holes in the web plate do not occupy more than 20 per cent of the web section area, they may be ignored (5.4.3.1).

For the buckling check, $V \leq V_{RS}$. V is due to the factored loading that arises in any bay between transverse stiffeners, while V_{RS} is for the same bay and is based on the ultimate failure by buckling. V_{RS} takes advantage of the post buckling behaviour and is as follows (5.4.3.3):

$V_{RS} = ([v_1 + v_{tf}]p_{vw}dt)/\gamma_m$ if there are no tongue plates or
$V_{RS} = ([\{(V_1 + v_{tf}\}p_{vw}dt] + p_{vt}A_{te})/\gamma_m$ with plates

where,
d is the depth of the web measured between the flanges, or to the tongue plate tips,
t is the unreduced web plate thickness,
v_1 is the initial shear buckling factor read from Figure 5.4, using $\epsilon = (150/p_v)^{0.5}$ and
v_{tf} is the tension field factor of Section 5.4.3.4 (5.4.3.3).

For the yielding check at any cross-section, V due to factored loading should not exceed V_{RS}, where:

$V_{RS} = (p_{vw}A_{we}/\gamma_m)$ if there are no tongue plates and
$V_{RS} = ([\{p_{vw}A_{we}\} + \{p_{vt}A_{te}\}]/\gamma_m)$ if there are tongue plates.

The effective areas are found using k_z times the true thickness in any HAZ region as in Sections 4.4.2 and 4.4.3 (5.4.3.2).

For tension field action, the following two types of web panel need to be identified:

1) Panels of Type 1 are able to sustain a tension field, e.g. an internal panel or a panel in an end bay, having an adequate end-post which complies with Section 5.4.5.6. These panels can develop further shear resistance after the initial onset of buckling and v_{tf} is as follows:

   $v_{tf} = v_2 + mv_3$ for an unwelded panel or

$v_{tf} = k_z(v_2 + mv_3)$ for a panel with edge welds

where:
m is the lesser of m_1 and m_2, shear buckling factors.
m_1 is found from Figure 5.7, $m_2 = ([4p_{of}S_f]/[p_{ow}d^2t])^{0.5}$,
v_2 is the shear buckling factor found from Figure 5.5 and
v_3 is the shear buckling factor found from Figure 5.6.

It is necessary to include the flange plate, the tongue plate and a thickness reduction to allow for HAZ softening and local buckling, but ignoring holes, when determining S_f. The plastic modulus of the whole flange should be suitably shared in obtaining S_f for each web if the girder has two or more webs (5.4.3.4).

2) Panels of Type 2, that are in an end bay lacking an adequate end-post. Normally there is almost no tension field and thus $v_{tf} = 0$ in Section 5.4.3.3 (5.4.3.4).

For the HAZ check on webs with longitudinal welds, $V \leq V_{RS}$ at any such weld, where

$V_{RS} = ([k_z p_{vw} It]/[A_e \bar{y} \gamma_m])$ where

$A_e\bar{y}$ is the first moment of the gross excluded area outside the weld, A_e is the section area and y′ the distance the centroid of the neutral axis to the section area (5.4.3.5).

10.4 Longitudinally and transversely stiffened girders (5.4.4)

A similar procedure is used to determine the moment resistance for longitudinally and transversely stiffened girders, as is used for girders with transverse stiffeners only. The procedure involves a buckling check to Section 5.4.2.3 and a yielding check to Section 5.4.2.2.

In the buckling check, each longitudinal stiffener is assumed to provide a line of support to the web, dividing the web into separate sub-panels, in respect of local buckling. Improved values of k_L can then be used for the sub-panels when determining the effective section of the girder. The improved values are obtained using the correct width and stress pattern for each sub-panel when determining β (5.4.4.1).

For shear resistance, the HAZ check (5.4.3.5) and the yielding check (5.4.3.2) are not affected by any longitudinal stiffeners. However, for the buckling check carried out in accordance with Section 5.4.3.3, v_l and v_{tf} are found as follows (5.4.4.2):

v_l is found from Figure 5.4, with
 d = the depth of the largest sub-panel not the full web depth,
v_{tf} is found using
 $v_{tf} = v_2 + mv_3$ for an unwelded panel or
 $v_{tf} = k_z(v_2 + mv_3)$ for a panel with edge welds as appropriate (5.4.3.4), with m, v_2 and v_3 obtained as follows (5.4.4.2):
 m is the lesser of m_1 and m_2. The values of m_1, v_2 and v_3 being calculated in accordance with Appendix K, using the formulae relating to Figures 5.7, 5.5 and 5.6 respectively. Figures 5.5, 5.6 and 5.7 are not to be used for web panels with longitudinal stiffeners. d is the depth of the web measured between the flanges, or to the tongue plate tips, and v_l is found as above.
 $m_2 = ([4p_{of}S_f]/[p_{ow}d^2t])^{0.5}$.

10.5 Web stiffeners and tongue plates (5.4.5)

Web stiffeners, as shown in Figure 5.3, are considered as single-sided or double-sided and are placed in three groups (5.4.5.1):

1) type A, are intermediate stiffeners; that is, a transverse stiffener other than those covered by type B,
2) type B, are bearing stiffeners; that is, a transverse stiffener at a point of concentrated load or at a point of reaction and
3) type C, longitudinal stiffeners, spanning longitudinally between transverse stiffeners.

The three groups of web stiffeners usually must comply with Section 5.4.5.2 for compactness (types A, B, C), with Section 5.4.5.4 for stiffness (types A, B, C) and with Section 5.4.5.5 for stability (types A, B).

Transverse stiffeners must extend unbroken from flange to flange. At a flange, provision must be made to transfer the applied force into the bearing stiffener, if it is fitted, but it is not necessary for the flange and the stiffener to be connected.

Longitudinal stiffeners should be continuous from one web bay to the next, or the separate lengths should abut onto the transverse stiffener between them (5.4.5.1).

Stiffeners should be of compact section, with respect to axial compression resistance (5.4.5.2).

The effective stiffener section consists of the actual stiffener or, if double-sided, the pair of stiffeners, plus an effective width of web plate b_e as shown in Figure 5.8. b_e extends b_l either side of the stiffener attachment or attachments. b_l is found as follows (5.4.5.3):

b_l = the lesser of 0.13a and $15\epsilon t$ for a transverse stiffener, unless the transverse stiffener is at the end of the girder when b_l on the outboard side becomes the lesser of a_o and $7\epsilon t$, where a_o is the distance from the stiffener to the free edge of the web plate, or

b_l = the lesser of $0.13d_{av}$ and $15\epsilon t$ for a longitudinal stiffener, where $\epsilon = (150/p_v)^{0.5}$ (p_v is the limiting stress for the web material as found in Tables 4.1 and 4.2) and d_{av} is the average depth of the two sub-panels lying either side of the longitudinal stiffener.

For a transverse stiffener, with $a/d \leq 2.5$, $I_s \geq dt^3([2d/a] - 0.7)$, but for a longitudinal stiffener, with $d_{av}/a \leq 2.5$, $I_s \geq at^3([2a/d_{av}] - 0.7)$, where I_s is the second moment of area of the full section of the effective stiffener about a centroidal axis parallel to the web.

The stiffness conditions above do not apply when the panel proportions are outside the above range (5.4.5.4).

When considering the stability of transverse stiffeners only, the effective stiffener section is assumed to be a strut carrying a thrust P under factored loading as follows:

$P = V/3$ for an intermediate stiffener and
$P = P_l + (V/3)$ for a bearing stiffener.

P must not exceed the factored axial resistance of the strut as determined in Section 4.7. Account must be taken of column buckling out of the plane of the web, local squashing, but not torsional buckling. For column buckling:

$l = d$ for $(a/d) \geq 1.5$ and also for any end stiffener, but
$l = (d/[1.6 - \{0.4a/d\}]^{0.5})$ for $a/d < 1.5$.

If a has different values on opposite sides of the stiffener, the average value is used.

Bending effects due to the eccentricity of P and the centroidal axis of the effective section, are considered using the interaction formulae given in Sections 4.8.3 and 4.8.4.4. M_y is the moment due to the action and $M_x = 0$ (5.4.5.5).

The tension field action can only be considered when determining the shear force resistance of an end bay of a plate girder if there is an adequate end-post at the outer end of the web panel (5.4.5.6). The end-post is designed to act as a bearing stiffener, resisting the reaction at the girder support, and to act as a short beam spanning between the girder flanges, resisting the tension field in the plane of the web.

When the end-post is acting as a short beam, it has to resist a shear force V_{ep} together with a moment M_{ep} in the plane of the web, where

$$V_{ep} = 0.6p_v dt(1 - v_1[\{\langle q/P_v\rangle - v_1\}/v_2])^{0.5} \text{ and}$$
$$M_{ep} = 0.1dV_{ep}$$

where v_1 and v_2 are end-panel factors found from Figures 5.4 and 5.5, or if longitudinally stiffened from Section 5.4.4.3, and q is the mean shear stress due to the end-panel of the web under factored loading, based on the unreduced thickness. It can be assumed, when calculating q, that part of the shear force on the girder is carried by any tongue plates that are fitted (5.4.5.6).

The end-post must be securely connected to both of the girder flanges and be one of the following forms:

1) inserted material, connected to the end of the web plate or
2) comprise two double-sided transverse stiffeners, which form the flanges of the short beam, together with a strip of web plate between them. One of the transverse stiffeners should fulfil the bearing role.

Some end-posts provide the only means of resisting twist at the end of a girder and the following must apply (5.4.5.7):

$$I_{ep} \geq (d^3 t_f R)/(250W).$$

A tongue plate is a plate that extends in from a flange to form a thickened outer part to the web. It is effective if classified as compact when considered as a plain outstand in axial compression as in Section 4.3.3 (5.4.5.8).

If the tongue plate is of two- or three-ply construction, that is, a web plate connected to an element or elements integral with the flange, t is the total thickness used for checking for tongue plate compactness. In bolted or riveted construction, any outstand beyond the last line of bolts or rivets must also be compact (5.4.5.8).

10.6 Use of corrugated or closely stiffened webs (5.4.6)

This section deals with girders where the transverse web reinforcement is corrugations or closely spaced stiffeners with a pitch of less than 0.3

times the depth between the flanges. This transverse reinforcement is considered sub-critical and that it will deform with the web in an overall buckling mode (5.4.6.1).

For a web consisting of applied stiffeners and a flat plate, the moment resistance is found using Section 5.4.2. If the web is corrugated, the flanges provide all the moment resistance (5.4.6.2).

V_{RS} is found as in Section 5.3.5 for multi-stiffened plating in shear (5.4.6.3).

10.7 Girders under combined moment and shear (5.4.7)

The moment–shear interaction for plate girders is shown in Figure 5.9a for bays unable to sustain a tension field and in Figure 5.9b for bays with a tension field. Further diagrams can be constructed for any bay between transverse stiffeners to determine M_{RSO} with coincident shear force V due to factored loading.

10.8 Design example 1

The plate girder shown in Figure 10.1 opposite consists of two flange plates 300 mm wide × 20 mm thick and a 15 mm thick web plate with a depth d = 810 mm. The distance between the transverse stiffeners is a = 1000 mm as shown in Figure 10.2 overleaf. The panel size is the same on opposite sides of the stiffener. There is no eccentricity of the line of action of any load P and the centroidal axis of the effective stiffener section.

Determine a suitable web stiffener design if the plate girder is to be compact. The flange, web and stiffener materials are to have no strength loss due to HAZ effects and to be of durability rating A.

Design example 1 at 1.9 in Chapter 1 gives a series of alloys that would be suitable. Of the non-heat-treatable alloys, 5083-O and 5083-F are available in the required plate thicknesses (Table 2.2). For the heat-treatable alloys 6082-T4 and 7020-T4 are available in the required thicknesses (Table 2.1). 5083-O and 5083-F both have durability ratings of A, while 6082-T4 has a durability rating of B, and 7020-T4 a durability rating of C. The choice is between 5083-O and 5083-F. In this case 5083-F will be used, giving $p_o = 130$ N/mm^2, $p_a = 170$ N/mm^2 and $p_v = 75$ N/mm^2 (Table 4.2).

Web stiffeners of Type A of Figure 5.3 of BS 8118: Pt 1 are being considered (5.4.5.1). In order that the predicted resistances can be achieved, the web stiffeners must comply with Section 5.4.5.2 for compactness, Section 5.4.5.4 for stiffness and Section 5.4.5.5 for

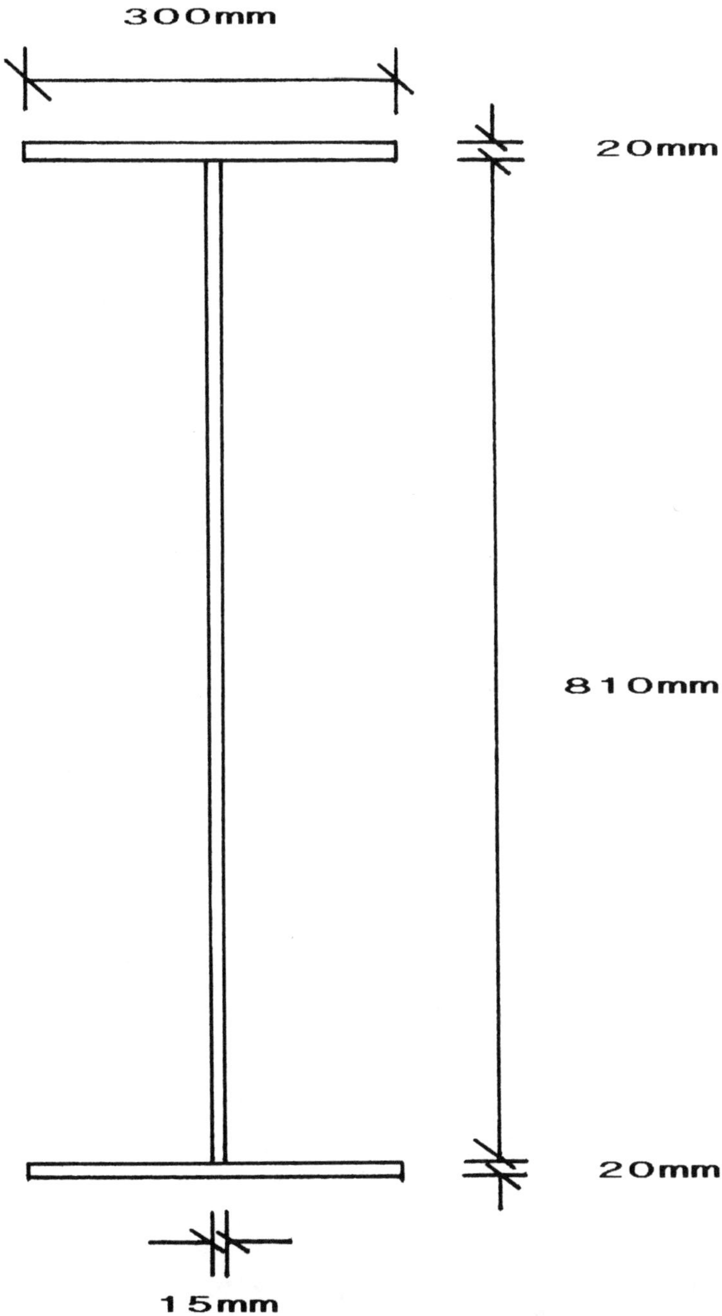

Figure 10.1 Section through the plate girder

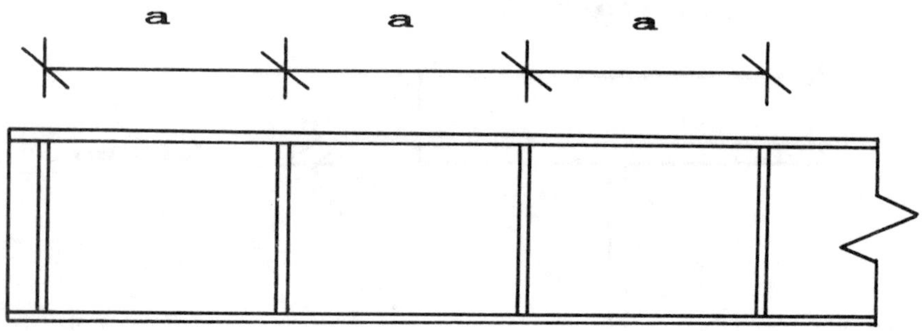

Figure 10.2 Elevation of the plate girder

stability. The web stiffener is a transverse stiffener and must extend from flange to flange without a break (5.4.5.1).

All stiffeners must be of compact section in terms of resistance to axial compression (5.4.5.2) and comply with Section 4.3.3.4.

For axial compressive resistance in compact sections $\beta \leq \beta_o$, which for welded outstand web stiffener elements gives

$$\beta_o = 6\epsilon$$
$$= 6 \times (250/130\ \text{N/mm}^2)^{0.5}$$
$$= 8.32 \text{ (Table 4.3).}$$

β = b/t (4.3.2.2a) and must be less than 8.32, therefore let t = 10 mm, b = 80 mm and have a stiffener on both sides of the web as shown in Figure 10.3 opposite.

To check the effective stiffener section, for a transverse stiffener, b_l = the lesser of 0.13a and 15ϵt (5.4.5.3 and Figure 5.8). a is the spacing between the transverse stiffeners, which is 1000 mm, and t = 10 mm,

$\epsilon = (150/75\ \text{N/mm}^2)^{0.5}$
$= 1.414$, giving
b_l = the lesser of 0.13×1000 mm = 130 mm or $15 \times 1.414 \times 10$ mm = 212.13 mm, giving b_l
= 130 mm.
b_e = 130 mm + 10 mm + 130 mm
= 270 mm (Figure 5.8) and shown in Figure 10.3 opposite.

For a transverse stiffener $I_s \geq dt^2([2d/a] - 0.7)$, provided $a/d \leq 2.5$ for the stiffener. d is the depth of the web between the flanges = 810 mm, giving

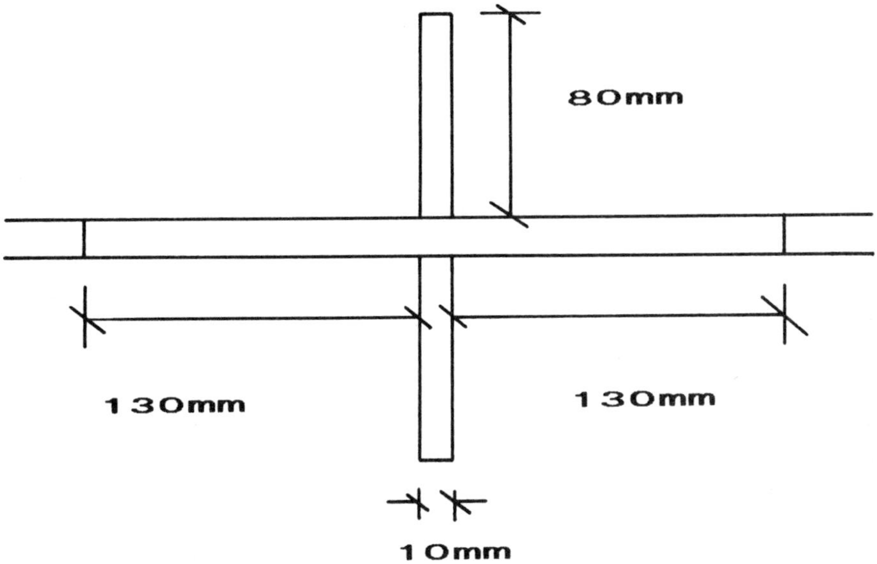

Figure 10.3 Effective stiffener section

$a/d = 1000\ mm/810\ mm$
$= 1.23 < 2.5$, therefore satisfactory (5.4.5.4).
$I_s \geq dt^2([2d/a] - 0.7)$
$\geq 810\ mm \times 10^3 mm^3([2 \times 810\ mm/1000\ mm] - 0.7)$
$\geq 745.2 \times 10^3 mm^4$.

But

$$I_s = ([270\ mm \times \{15\ mm\}^3/12] + 2 \times [\{\langle 10\ mm \times 80^3 mm^3\rangle/12\} + \{80\ mm \times 10\ mm \times \langle 40\ mm + 7.5\ mm\rangle^2\}])$$
$$= 4.539 \times 10^6 mm^4$$

which is greater than $dt^2([2d/a] - 0.7)$. Therefore the transverse web stiffener is satisfactory and could be reduced in size.

For stability, the thrust in the stiffener under factored loading is P. For an intermediate stiffener, $P = (V/3)$ where V is the average value of the shear force arising in the web panels on either side of the stiffener (5.4.5.5).

For column buckling with a/d (1000 mm/810 mm) ≤ 1.5,

$$l = d/(1.6 - [\{0.4a\}/d])^{0.5}$$
$$= 810\ mm/(1.6 - [\{0.4 \times 1000\ mm\}/810\ mm])^2$$

= 770 mm, the effective strut length.

$P \leq P_R$, the factored axial resistance of the strut. Using Section 4.7, check for out-of-plane of the web column buckling and local squashing. There is no eccentricity between the line of action of P and the centroidal axis of the effective section.

For the column buckling check, P must not exceed P_R, where $P_R = p_s A/\gamma_m$ (4.7.3). A is related to the effective stiffener section (Figure 5.8). p_s is found from Figure 4.10 (4.7.4),

$\lambda = l/r$ where
$r = (I/A)^{0.5}$
$= ([4.539 \times 10^6 \text{ mm}]/[\{2 \times 10 \text{ mm} \times 80 \text{ mm}\} + \{270 \text{ mm} \times 15 \text{ mm}\}])^{0.5}$
$= 28.34$ mm.
$\lambda = l/r$
$= 770 \text{ mm}/28.34 \text{ mm}$
$= 27.17$.

For a welded, symmetric section Figure 4.10b is used (4.7.6.1 and Table 4.10). Also, as there are no HAZ effects to be considered as ($k_z = 1$), $p_l = p_o = 130 \text{ N/mm}^2$ (4.7.6.2). Moving up the y axis on Fig 4.10b to 130 N/mm^2 and then moving to the right between the 120 N/mm^2 and the 140 N/mm^2 curves to $\lambda = 27.17$ gives $p_s = 119 \text{ N/mm}^2$. $\gamma = 1.2$ (Table 3.3), giving

$P_R = (119 \text{ N/mm}^2 \times 5\,650 \text{ mm}^2)/1.2$
$= 560.29$ kN.

For local squashing (4.7.7), a compact section with no HAZ effects and no holes

$P_{RS} = p_a A_n/\gamma_m$
$= p_a A/\gamma_m$
$= (170 \text{ N/mm}^2 \times 5\,650 \text{ mm}^2)/1.2$
$= 800.42$ kN.

$P_{RS} > P_R$, therefore the limiting value is $P_R = 560.29$ kN.
$p \leq P_R$ (560.29 kN) and $P = (V/3)$, giving the maximum value of V as 1681 kN.

The shear resistance of a transversely stiffened girder V_{RS} requires a buckling check and a yielding check (5.4.3.1).

For the buckling check (5.4.3.3), $V \leq V_{RS}$ where

$$V_{RS} = (v_1 + v_{tf})p_{vw}dt/\gamma_m$$
$$= (v_1 + v_{tf}) \times 75\ N/mm^2 \times 810\ mm \times 15\ mm/1.2$$

v_1 is found from Figure 5.4.

$a/d = 1000\ mm/810\ mm$
$= 1.23$

$\epsilon = (150/75\ N/mm^2)^{0.5}$, giving

$d/(t\epsilon = 810\ mm/(15\ mm \times 1.414)$
$= 38.18$

Thus from Figure 5.4 $v_1 = 1.0$

v_{tf} is the tension field action (5.4.3.4).

For a Type 1 internal panel with edge welds $v_{tf} = k_z(v_2 + mv_3)$, with m being the lesser of m_1 and m_2. $k_z = 1.0$ (Table 4.5).

v_2 is found from Figure 5.5, $a/d = 1.23$, $d/(t\epsilon) = 38.18$ giving $v_2 = 0.0$

v_3 is found from Figure 5.6, $a/d = 1.23$, $d/(t\epsilon) = 38.18$ giving $v_3 = 0.0$

m_1 is found from Figure 5.7, $a/d = 1.23$, $d/(t\epsilon) = 38.18$ giving $m_1 = 0.0$

$m_2 = ([4p_{of}S_f]/[p_{ow}d^2t])^{0.5}$
$= ([4 \times 130\ N/mm^2 \times 300\ mm \times \{20\ mm\}^2]/(130\ N/mm^2 \times 810\ mm^2 \times 15\ mm \times 4])^{0.5}$
$= 0.1104$

$m_1\ (0.0) < m_2\ (0.1104)$, thus use m_1 giving

$v_{tf} = 1.0(0.0 + [0.0 \times 0.0])$
$= 0.0$

Thus

$$V_{RS} = (1.0 + 0.0) \times 75\ N/mm^2 \times 810\ mm \times 15\ mm/1.2$$
$$= 759.38\ kN.$$

For the yielding check $V \leq V_{RS}$, where $V_{RS} = p_{vw}A_{we}/\gamma_m$ as there are no tongue plates (5.4.3.2). $p_{vw} = p_v = 75\ N/mm^2$ (Table 4.1), $A_{we} = 810\ mm \times 15\ mm$, $\gamma_m = 1.2$ giving

$$V_{RS} = (75\ N/mm^2 \times 810\ mm \times 15\ mm)/1.2$$
$$= 759.38\ kN.$$

Both the yielding check and the buckling check give $V_{RS} = 759.38$ kN. $V \leq V_{RS}$ (759.38 kN) and $P \leq P_R$ (560.29 kN), also $P = (V/3)$. Therefore $V = 759.38$ kN is the limiting value, giving $P = (759.38\ kN/3) = 253.13$ kN which is satisfactory.

10.9 Design example 2

Determine the moment resistance of the transversely stiffened plate girder described in 10.8, 'Design example 1'.

To determine the factored moment of resistance, a buckling check and a yielding check are required (5.4.2.1).

For the web buckling check, the web is considered as unwelded and slender, with only transverse stiffeners and no longitudinal stiffeners (5.4.2.3). HAZ effects caused by welding on the transverse stiffener are ignored.

$M_{RS} = p_o Z_e/\gamma_m$ or $M_{RS} = p_o Z_n/\gamma_m$, whichever is the smaller (4.5.2.2e). Local buckling and HAZ effects are to be allowed for. Holes are ignored (5.4.2.3). There are no holes, therefore $Z_n = Z$ (4.5.2.3a). For 5083-F, $k_z = 1.0$ (Table 4.5), therefore $Z_e \leq Z_n$.

The compressive edge and the tension edge of the web are the same distance from the neutral axis as the beam is assumed to be in pure bending. k_L, the effective thickness factor, is found using Section 4.3.4 and β is found using Section 4.3.2.2 for web buckling.

For the flat welded outstand element (flange), use Figure 4.5 Curve B (Table 4.4). For the welded flat internal element (web), use Figure 4.5 Curve D (Table 4.4).

For the unreinforced flange

$$\beta = b/t$$
$$= 142.5\ \text{mm}/20\ \text{mm}$$
$$= 7.125\ (4.3.2.2a).$$

For the web

$$\beta = 0.35d/t$$
$$= 0.35 \times 810\ \text{mm}/15\ \text{mm}$$
$$= 18.9\ (4.3.2.2b).$$

$$\epsilon = (250/130\ \text{N/mm}^2)^{0.5}$$
$$= 1.387, \text{ giving}$$
$\beta/\epsilon = 5.138$ for the flanges and $k_L = 1.0$ (Figure 4.5) and
$\beta/\epsilon = 13.629$ for the web and $k_L = 1.0$ (Figure 4.5).

As both k_z and $k_L = 1.0$, the thickness of the web and the flange are not reduced giving $Z_e = Z_n = Z = 6.426 \times 10^6 \text{mm}^3$ and

$$M_{RS} = (130\ \text{N/mm}^2 \times 6.426 \times 10^6 \text{mm}^3)/1.2$$
$$= 696.24\ \text{kN-m}.$$

For the yielding check M_{RS} is found treating the section as semi-compact with no holes or HAZ effects ($k_z = 1$) and welded (5.4.2.2). This gives

$M_{RS} = p_o Z_{ne}/\gamma_m$ (4.5.2.2d),

where:
p_o = 130 N/mm^2,
γ_m = 1.2.
Z_{ne} = Z as there are no holes and k_z = 1.0.
Z_{ne} = I/425 mm
= ([{15 mm × ⟨810 mm⟩^{3}}/12] + 2[{⟨300 mm × ⟦20 mm⟧^{3}⟩/12} + {⟨300 mm × 20 mm⟩ × ⟨415 mm⟩^{2}}])/425 mm
= 6.427 × 10^{6}mm^3 giving
M_{RS} = (130 N/mm^2 × 6.427 × 10^{6}mm^3)/1.2
= 696.24 kN-m.

In this case Appendix E has not been used (5.4.2.2).

Both the yielding check and the buckling check give M_{RS} = 696.24 kN−m.

11 Static design of joints (6)

11.1 General (6.1)

The term joints includes (6.1):

1) adhesives,
2) fasteners, which includes black bolts, close tolerance bolts, high strength friction grip bolts (HSFG bolts), pins, rivets and special fasteners and
3) welding, including butt and fillet welds.

The manufacturers of cast or forged elements will need to be consulted when jointing those elements.

Typical joints include connections between:

1) structural members such as beam to column,
2) the elements of a built-up member, such as flanges to webs, splices and
3) connections between localized details and structural members, such as beam to bracket or lug and clevis in a tension member.

Joints must be designed to satisfy the fatigue and static strength limit states. A serviceability limit state check is not required. The joint factored loading is calculated using the load factors given in Section 3. A serviceability limit state check is required, however, for pin joints that are frequently assembled and disassembled, for joints where deflections are critical and for friction grip bolts where there must be no slip.

Where there is load reversal, a fastener should be either close tolerance bolts, HSFG bolts, solid rivets, special fasteners or turned barrel bolts, that prevent movement.

Fasteners not complying with the British Standards can only be used with the agreement of the designer.

11.2 Riveted and bolted joints (6.2)

11.2.1 Design considerations

For bolted and riveted joints under factored loading, the action of the load at any fastener position must not exceed the factored resistance of the fastener at that point (6.2.1).

Where a fastener consists of groups of bolts, rivets or special fasteners, the design must make realistic assumptions about the distribution of internal forces regarding relative stiffness (6.2.2).

Except for rivets in compression, the design of plies at sections containing holes uses the minimum net area. For some friction grip bolted joints, the design of the plies is based on the minimum gross area (6.2.3).

When a joint length, in the direction of load transmission, measured between the centres of the end fasteners, is more than 15 times the nominal fastener diameter (d_f), or when there are more than five fasteners in that direction, there is a reduction in the average strength of the individual fastener (6.2.4).

11.2.2 Geometrical and other considerations (6.3)

The minimum spacing between bolt or rivet centres is 2.5 times the bolt or rivet diameter, except for HSFG bolts, where the limit can be a limit state, the sizes of the bolt head, the spanner or the washer (6.3.1).

The following recommendations can only be used for cover plate joints and lap joints between flat plates where the thicknesses are similar. For tension members, the maximum bolt or rivet spacing, in the direction of the stress, is 16t or 200 mm, where t is the thickness of the thinnest outside ply. For compression or shear members, the maximum bolt or rivet spacing is 8t or 200 mm. Further, the maximum spacing of adjacent bolts or rivets on a line adjacent and parallel to an edge of an outside ply is 8t or 100 mm. The spacing may be increased by 50 per cent if the bolts or rivets are staggered on adjacent lines, with the lines being not more than 75 mm apart. The maximum spacing of adjacent bolts and rivets for tension members is 32t or 300 mm. The maximum spacing of adjacent bolts and rivets is 20t or 300 mm for compression or shear members. For other joints, the local geometry and the loading must be considered when obtaining bolt and rivet spacings (6.3.2).

For extruded, machined or rolled edges, the minimum edge distance to the centre of a bolt or a rivet is 1.5 times the bolt or rivet diameter. For edge distances, on the bearing side, of between 1.5 and 2 times the bolt or rivet diameter, the bearing capacity is reduced in accordance with Section 6.4.4. The edge distance is increased by 3 mm for sheared edges (6.3.3).

Table 3.1 of BS 8118: Pt 2 gives hole clearances. Bolts used for transmitting fluctuating loads, but not wind loads, must be close-fitting or HSFG (6.3.4). Section 2.3 of BS 8118: Pt 2 should be consulted regarding washers. Locking devices should be used on nuts liable to work loose due to stress fluctuation or vibration (6.3.8).

Rivet grip length should not exceed five times the hole diameter (6.3.7).

If the packing thickness exceeds 25 per cent of the fastener diameter, or 50 per cent of the ply thickness, the factored design resistance must be reduced if the fastener is carrying shear through the packing (6.3.5).

When calculating the length in bearing, 50 per cent of any countersinking depth of a bolt or rivet is neglected, unless the bolt or rivet is in shear when no reduction is necessary. When in axial tension, the factored design resistance of a countersunk bolt or rivet is two-thirds that of the plain bolt or rivet of the same diameter. The maximum countersunk depth is 4 mm less than the thickness of the countersunk part, unless proved satisfactory by testing (6.3.6).

At intersections, the centroidal axes of members at the joint should meet at a single point. The setting-out lines of the bolts can be used instead of the centroidal axes for the bolted framing of angles and tees (6.3.9).

11.3 Factored resistance of individual bolts and rivets (6.4)

HSFG bolts complying with BS 4395: Pt 1 are dealt with in Section 6.5. p_f, the limiting stress for solid bolts and rivets, is defined as (6.4.1):

1) p_f = the guaranteed minimum yield stress for the bolt or rivet stock in steel fasteners,
2) p_f = the lower of $0.5(f_{0.2} + f_u)$ and $1.2f_{0.2}$ for stainless steel bolts and stainless steel rivets and

3) p_f = the values given in Table 6.1 for the aluminium alloy bolts and rivets given in Table 2.3. The shear strength value, if available from tests on bolts or from tests on rivets in the as-driven condition, may be used, with α_s in V_{RS} in Section 6.4.2 reduced to 0.33 from 0.6.

For a single bolt or rivet in single shear (6.4.2):

$$V_{RS} = (\alpha_s p_f A_{es} K_1)/\gamma_m$$

where:
α_s = 0.6 for aluminium bolts and rivets and
α_s = 0.7 for steel bolts or rivets.
γ_m = the material factor, equal to 1.2 for all aluminium, stainless steel and steel bolts and rivets.

For bolts,
A_{es} = A_{tb}, which is the stress area of the threaded part of the bolt, when the shear plane passes through that area or
A_{es} = A_{SH}, the area of the shank, when the shear plane passes through the shank.

For rivets,
A_{es} = A_h, which is the hole area, with
K_1 = 1.0 for rivets,
K_1 = 0.95 for close tolerance bolts and
K_1 = 0.85 for normal clearance bolts.

Aluminium rivets should not be used in tension. For a single fastener in axial tension, the factored resistance is (6.4.3):

$$P_{RT} = (a p_f A_{tb})/\gamma_m$$

with
a = 1.0 for stainless steel and steel bolts and rivets and
a = 0.6 for aluminium bolts.

For a bolt or rivet in bearing, the effective factored resistance is the lesser of B_{RF}, the single fastener factored resistance in bearing, and B_{RP}, the bearing capacity of the connecting ply (6.4.4), where:

$B_{RF} = ([d_f t 2 p_f]/\gamma_m)$ and
B_{RP} is the lesser of $([c d_f t p_a]/\gamma_m)$ or $(e t p_a/\gamma_m)$.

$c = 2$ when $(d_f/t) < 10$,
$c = (20t/d_f)$ when $10 < (d_f/t) < 13$ and

c = 1.5 when $(d_f/t) > 13$.
e = the distance from the adjacent edge in the direction that the fastener bears to the centre of the hole.
p_a is the lesser of $0.5(f_{0.2} + f_u)$ and $1.2f_{0.2}$ for the connected ply material.

Except for aluminium rivets, bolts and rivets subjected to both shear and tension must also satisfy the following (6.4.5):

$$(P/P_{RT})^2 + (V/V_{RS})^2 \leq 1$$

11.4 HSFG bolts (6.5)

For aluminium structures, only the pre-loaded general grade HSFG bolts to BS 4395: Pt 1 can be used (6.5.1).

Where the proof strength of the connecting parts exceeds 230 N/mm^2, the joints may be designed by calculation. If the proof strength of the connecting parts is less than 230 N/mm^2, the joint strength using general grade HSFG bolts has to be proved by testing. The relaxation of bolt pre-load, due to tension, has to be considered. A temperature increase increases friction capacity, bolt tensile stress and bearing stress under the washers. These changes must also be considered (6.5.1).

The ultimate limit state of static strength for HSFG bolts in normal clearance holes, as given in Table 3.1 of BS 8118: Pt 2, is the lesser of the bearing capacity of Section 6.4.4 or the shear capacity of Section 6.4.2 (6.5.2).

For the serviceability limit state deformation check of a connection using HSFG bolts, $\gamma_s = 1.2$. The limit state is reached when the shear load on any bolt equals the bolt's friction capacity (6.5.3). The friction capacity of an HSFG bolt is (6.5.4):

$$F_c = ([P_p\mu_s N_F]/\gamma_s),$$

where:
N_F is the number of friction interfaces,
P_p is the pre-stress load,
$\gamma_s = 1.1$ if μ_s is found from tests, but $= 1.3$ if μ_s is taken as 0.33.
μ_s is the slip factor.

For HSFG bolts, the pre-stress load P_p is as follows (6.5.5):

$$P_p = P_o - 0.9S_{tb}$$

where:
P_o is the bolt proof load and
S_{tb} is any applied external load in the axial direction of the bolt.

The value of the slip factor μ_s may be assumed to be 0.33 if (6.5.6):

1) all the connected parts are of aluminium alloy,
2) the friction interfaces have been treated by blasting with aluminium oxide G38 grit complying with BS 2451,
3) the total thickness of the connected parts exceeds the bolt diameter and
4) the gross area stress in the parts do not exceed $0.6f_{0.2}$.

If any of the four conditions are not met, μ_s is found from tests to BS 4604: Pt 1.

Often the number of bolts required to satisfy the friction capacity, with $\mu_s = 0.33$, is larger than the number required to satisfy the ultimate limit state of static strength. In this case it may be necessary to treat the interfaces to increase the slip factor (6.5.6).

11.5 Pinned joints (6.6)

A pinned joint consists of a single pin not loaded in single shear, nor loaded axially. The pin does not clamp the connected parts and one of the members to be joined must have a clevis or a fork end. The system used to retain the pin must be able to resist a lateral load of 10 per cent of the pin's total shear load (6.6.1).

For solid pins, the bending stresses are calculated using the distance between the centres of the bearings as the effective span. Further, if half the pin diameter is less than the bearing plate thickness, then the effective span may be altered due to the variation of bearing pressure across the plate thickness.

Pins that have been re-inserted need to be checked for the serviceability limit associated with the limit of elastic behaviour. Under factored load, the mean shear stress in the pin should not exceed $(0.6p_a/\gamma_m)$ and the bending stress in the pin should not exceed $(1.2p_a/\gamma_m)$.

For permanently installed pins, a fully plastic distribution of bending stress is assumed at the factored design load (6.6.2).

It is assumed that the following rules are only used when the line of action of the load is in the same direction as the grain flow in the connected parts (6.6.3).

For a pin-connected tension member, the net area across the pin hole, normal to the axis, must be equal to or greater than ($[1.33P\gamma_m]/p_a$). The thickness of the connected member must be at least:

($P\gamma_m/[1.6p_a d_f]$) for permanent installations and
($P\gamma_m/[1.4p_a d_f]$) for demountable installations.

On either side of the axis of the member, the net area of any section, measured at 45° or less to the member axis, must be equal to or greater than ($0.9P\gamma_m/p_a$). Eight times the bearing plate thickness must be greater than the net width of the bearing plate at the pin hole, measured normal to the member axis. The pin hole diameter must be not more than 5 per cent greater than the pin diameter. Pin plates, and any connections between them and the member, must be designed to carry a share of the total axial load, in proportion to the plate's share of the total pin bearing area.

11.6 Welded joints (6.7)

11.6.1 General

Welds must be in accordance with Section 3.9 of BS 8118: Pt 2 and use the recommended filler and parent material combinations given in Table 2.8 of Pt 1. It is also necessary to consider the following points when selecting a joint type (6.7.1):

1) suitable detailing can reduce stress concentrations,
2) suitable detailing can facilitate good welds and allow proper inspection,
3) suitable detailing can reduce corrosion (Section 4.3 in BS 8118: Pt 2),
4) the fatigue strength of the member is affected by the joint,
5) the effects of welding distortion and
6) how the static strength of the member is affected by the joint.

Close to the weld, the static strength of the parent metal may be changed (Section 4). Usually, welding reduces the strength in the following alloys:

1) 5*** series in the non-heat-treatable alloys in any work-hardened condition,
2) 6*** series in heat-treatable alloys in most heat-treated conditions and
3) 7*** series in heat-treatable alloys in most heat-treated conditions.

HAZ effects in the non-heat-treatable alloys in the F or O conditions can be ignored. There are exceptions as given in Table 4.5, when $k_z = 1$.

If the material suffers from a strength reduction due to welding, the weld should be parallel to the direction of the applied load. If welds are transverse to the applied load, they should be placed in areas of low stress. This also applies to welded attachments (6.7.2).

Joint fatigue strength is affected by both the local weld geometry as well as the severity of the stress concentration. Section 7.3 gives information on the fatigue classifications of some joint details that may be used to determine the most appropriate detail to increase fatigue resistance (6.7.3).

Joint details should ensure no inaccessible crevices or pockets that could increase corrosion. Any cavities should be accessible for inspection, sealed by protective compound or sealed by welding (6.7.4).

For welded joints, edge preparations are given in BS 3019: Pt 1 and BS 3571: Pt 1 and include butt welds, fillet welds and the use of permanent or temporary backing bars. BS 499: Pt 1 defines the welding positions (6.7.5).

The welding of aluminium causes distortion and shrinkage. This distortion and shrinkage must be balanced or compensated to keep the dimensions and shape of the completed structure as designed. The designer and the fabricator should consult together to minimize the problem (6.7.6).

For every weld, the following drawings and specifications need to be provided (6.7.7):

1) edge preparation and welding position;
2) filler and parent material;
3) quality control requirements for:
   a) acceptance levels for weld quality
   b) inspection levels of welded joints
   c) weld procedure approval
   d) weld quality class
   e) weld repair procedure and
   f) welder approval;
4) special requirements, e.g. pre-heat temperatures, interpass temperatures and weld profile smoothness;
5) weld dimensions and
6) the welding process.

'Normal' weld quality is assumed, unless otherwise specified. For both fatigue and static resistance, if the actions due to factored loading are equal to or less than one-third of the factored resistance of the joint or of the member, a lower degree and quality of inspection is acceptable, but a 'minimum' quality level may be specified. If the joints are designed to fatigue strength requirements, Section 7.8.5 should be consulted (6.7.7).

11.6.2 Butt welds (6.7.8)

Neither bending moments about the longitudinal axis of the weld, nor tensile forces, should be transmitted through intermittent and single-sided partial penetration butt welds. For a partial penetration butt weld, the effective throat thickness t_e, shown in Figures 6.1b and 6.1c (BS 8118: Pt 1), is taken as:

1) the depth of the weld preparation, where this is of the J or U type, or
2) the depth of the weld preparation, less either 3 mm or 25 per cent (whichever is the less), where this is of the bevel type or the V type.

Procedure trials may also be used to determine the throat thickness. The throat thickness must not be greater than the penetration consistently achieved, ignoring weld reinforcement. If a backing plate is used with a single-sided butt weld, full penetration may be assumed. A superimposed fillet weld may be taken into account in a tee joint.

11.6.3 Fillet welds (6.7.9)

Moments should not be transmitted about the axis of single-sided fillet welds. If intermittent fillet welds are used, then the distance between the ends of adjacent welds, whether in line or staggered on alternate sides of the part, must be equal to or less than the following:

1) if in compression or shear:
   a) 10 times the thickness of the thinner parent metal or
   b) 300 mm and
2) if in tension:
   a) 24 times the thickness or
   b) 300 mm.

There should be a weld at each end of the part connected, in a line of intermittent welds.

The design resistance of a fillet weld joint is given in Section 6.9.2. At the end or at the side of a part, a fillet weld should continue around the corner for a length of at least twice the weld leg length. Overlapping HAZs must be considered (4.4.3.6). If, at a lap-jointed end connection, two longitudinal fillet welds alone are used, then the length of each weld must be at least equal to or greater than the distance between the welds.

Figure 6.2 (BS 8118: Pt 1) gives the effective fillet weld throat length g_t, the weld leg length g_l and the penetration p_t. g_t is the height of the triangle inscribed within the weld and measured perpendicular to the outer side of the weld. If the penetration can be consistently achieved, then the specified penetration p_t can be included in the fillet weld throat. If the procedure trials show that the necessary penetration beyond the root can be consistently achieved, then a large throat may be assumed.

The fillet weld throat dimension g_t, times the effective fillet weld length, gives the effective area of the fillet weld. If the fillet weld is in a hole or a slot, then the effective area must not be greater than the hole area or slot area.

11.7 Design strength of welded joints (6.8)

The strength of the material in the HAZ next to the weld fusion boundary, and the strength of the weld metal, must be considered in the design of welded joints. Section 6.9 has the limiting stresses for the HAZ material. If the factored weld resistance is greater than the adjacent material in the HAZ, then the deformation capacity of the joint is improved (6.8.1).

A group of welds that form a joint is designed assuming a realistic distribution of forces amongst the welds, related to the relative stiffnesses of the welds. Equilibrium with the external factored loads must be maintained (6.8.2).

In welded construction, the filler wire is chosen in accordance with Section 2.5.3.2 and Table 2.8. The values of the weld metal limiting stress p_w, for the permitted combinations of filler and parent metals shown in Table 2.8, are given in Table 6.2. When using Table 6.2 for dissimilar metals, the lower of the limiting stresses for the weld metals is used. Reference to Appendix D is required for particular filler materials and weld metal that may have a higher limiting stress. Experimental tests are required to determine limiting stresses for welds on parent metals not included in Table 6.2 or Appendix D (6.8.3).

Table 6.3 gives the limiting direct stress p_{az}, and limiting shear stress p_{vz}, for material in the HAZ.

11.8 Factored resistance of welds (6.9)

For a butt weld subjected to axial and shear loading:

$$(\sigma_1^2 + 3\tau_2^2)^{0.5} \leq (p_w/\gamma_m)$$

where:

σ_1 is the normal stress perpendicular to the throat section under factored loading and

τ_2 is the shear stress acting on the throat section parallel to the axis of the weld under factored loading.

For a butt weld subject to an oblique tensile load, shown in Figure 6.4 (BS 8118: Pt 1), then:

$$P_{RB} = (p_w l_e t_e [1 + 2\cos^2\theta]^{-0.5})/\gamma_m.$$

l_e may be omitted when P_{RB} per unit length is being found for a weld subject to in-plane bending.

l_e is the total weld length, when run-on and run-off plates are used. Without run-on and run-off plates the weld length is the total length less twice the weld width. The possibility of stress concentrations must be investigated when there are different thicknesses of parent metal on either side of the weld. Except where buckling can occur, the weld metal design stress in compression and in tension are equal (6.9.1).

If the line of the butt weld is perpendicular to the line of the action of the external load, and the joint is subject to no external shear forces, then:

$$P_{RB} = ([p_w l_e t_e]/\gamma_m).$$

If the external shear load is parallel to the line of the butt weld, then (6.9.1):

$$P_{RB} = ([p_w l_e t_e)/(3^{0.5}\gamma_m]).$$

For a fillet weld (6.9.2):

$$(\sigma_1^2 + 3[\tau_1^2 + \tau_2^2])^{0.5} \leq 0.85 p_w/\gamma_m$$

where τ_2 is the shear stress acting on the throat section perpendicular to the weld axis.

As shown in Figure 6.5 (BS 8118: Pt 1), the direction of the external loading, S, affects the values of σ_1, τ_1 and τ_2. For a simple transverse weld, where the load is applied perpendicular to the weld length, then:

$$P_{RF} = (0.85 p_w l_e g_t)/(2^{0.5}\gamma_m).$$

For a simple longitudinal fillet weld, where the load is applied parallel to the weld length, then:

$P_{RF} = (0.85p_w l_f g_t)/(3^{0.5}\gamma_m)$.

l_f is influenced by L and Figure 6.6 (BS 8118: Pt 1) gives the ratio of (l_f/L) for (L/g_t) $\leq$ 50.

When the weld stress distribution and the adjacent parent material stress distribution match, l_e is as for butt welds. Further, if the weld is subjected to in-plane bending, P_{RF} per unit length is found by omitting l_e or l_f from the equations for P_{RF}.

For a HAZ next to a weld, P_{RFB} is as follows (6.9.3):

1) for a direct tensile force normal to the failure plane as shown in Figure 6.3 (BS 8118: Pt 1):

   P_{RFB} = ($[p_{az}Lt_e]/\gamma_m$) at the fusion boundary of a butt weld,
   P_{RTB} = ($[p_{az}Lt]/\gamma_m$) at the toe of a butt weld,

   P_{RFF} = ($[p_{az}Lg_1]/\gamma_m$) at the fusion boundary of a fillet weld and
   P_{RTF} = ($[p_{az}Lt]/\gamma_m$) at the toe of a fillet weld.

2) for a shear force in the failure plane as shown in Figure 6.3:

   V_{RFB} = ($[p_{vz}Lt]/\gamma_m$) at the fusion boundary of a butt weld,
   V_{RTB} = ($[p_{vz}Lt]/\gamma_m$) at the toe of a butt weld,

   V_{RFF} = ($[p_{vz}Lg_1]/\gamma_m$) at the fusion boundary of a fillet weld and
   V_{RTF} = ($[p_{vz}Lt]/\gamma_m$) at the toe of a fillet weld.

When checking P_{RTF} and V_{RTF}, a smaller value of t should be used (Figure 4.6(i) of BS 8118: Pt 1 and 4.4.3.1), because for thicker sections the HAZ does not extend the full thickness.

For combined direct force and shear loading on the HAZ, then:

$$(S_a/P_{RZ})^2 + (S_b/V_{RZ})^2 \leq 1$$

The factored resistance per unit length for a failure plane subjected to in-plane bending can be found by omitting L from the earlier equations. The factored resistance per unit length for a failure plane subjected to in-plane bending and shear must be reduced to allow for the combined effects of direct stress and shear (6.9.3c).

11.9 Bonded joints (6.10)

Adhesive bonding may be used to make structural joints for carrying shear loads in aluminium, but care needs to be exercised (BS 8118: Pt 2). For example, bonded joints should not be used in tension or where the forces will tend to open the joint. The bonded area should be as large as possible, with wide joints being better than long joints, except for very short joints. By reducing peel and cleavage stresses and stress concentrations at the end of laps, the performance of large bonded joints can be improved. Tapering off the ends of lap joints and introducing compensation pieces can be beneficial. Air pockets in the bonded joint should be avoided, also the bonded joint needs adequate time and support to allow the adhesive's strength to develop. The designer will need specialist advice on the suitability of the adhesive, plus further advice at all stages of the design and construction. This advice must be strictly followed (6.10.1).

For a bonded joint P_{RG} is influenced by:

1) assembly and curing procedures,
2) the direction of the stresses in the joint,
3) the design life,
4) the service temperature and environment,
5) the size and shape of the components to be joined,
6) the surface preparation procedures before bonding and
7) the thickness of the glue line.

Either validated test data or the testing of joints identical as far as possible to those to be used in service should be used to establish the joint strength. A minimum of five tests is need to determine P_{RG}.

$$P_{RG} = (R_m - 2S_d)/\gamma_m$$

where γ_m may need to be increased when operating in extremes of environment and temperature (6.10.2).

Manufacturer's test data may be used to give optimistic values for the initial design (Figure 6.7 of BS 8118: Pt 1). If only R_m is given, then $S_d = 0.1R_m$, but see also Appendix B. Thin sheet lap tests (BS 5350: Pt C5) may also be used, but will be conservative if used for structural design calculations (Figure 6.8 of BS 8118: Pt 1) (6.10.3).

11.10 Design example 1

Determine the factored resistance P_{RFB} and P_{RTB} of a butt weld in the HAZ in alloy 6063-T5. The loading is tensile, normal to the failure plane with L = 1000 mm. From Table 3.3, $\gamma_m = 1.3$,

a) at the fusion boundary F nearby an in-line butt weld (Figure 6.3a of BS 8118: Pt 1) with $t_e = t = 18$ mm (Figures 6.3 and 6.1 of BS 8118: Pt 1) and
b) at the toe T next to a tee-butt weld (Figure 6.3c of BS 8118: Pt 1) with in this case $t = 20$ mm.

For a butt weld at the HAZ (fusion boundary) F (Figure 6.3c of BS 8118: Pt 1):

$p_{az} = 95$ N/mm^2 (Table 6.3 of BS 8118: Pt 1).
$P_{RFB} = (p_{az}Lt_e)/\gamma_m$
$= 95$ N/mm$^2 \times 1000$ mm $\times$ 18 mm/1.3
$= 1315.38$ kN.

For a butt weld at the HAZ (toe) T (Figure 6.3c of BS 8118: Pt 1):

$p_{az} = 95$ N/mm^2 (Table 6.3 of BS 8118: Pt 1).
$P_{RTB} = (p_{az}Lt_e)/\gamma_m$
$= 95$ N/mm$^2 \times 1000$ mm $\times$ 20 mm/1.3
$= 1461.54$ kN.

11.11 Design example 2

A welded I-beam flange/web connection is to be fabricated using material 6063-T5, 20 mm thick. The weld will be intermittent and on both sides of the web but not staggered. The weld will be 1000 mm long with a spacing of 200 mm between welds. Run-on and run-off plates will be used. The weld throat thickness g_t is 7 mm and the minimum factored shear force required to generate the required bending moment is 750 kN. Determine a suitable weld material and say whether the weld has the required strength. The strength of the material in the HAZ next to the weld fusion boundary is not being considered.

The weld must be made in accordance with Section 3.9 of BS 8118: Pt 2, using the recommended combination of parent and filler materials given in Table 2.8 (6.7.1).

Alloy 6063-T5 suffers a strength reduction due to welding ($k_z = 0.75$, Table 4.5). The weld should be parallel to the direction of the applied load (6.7.2).

The fillet weld will be double-sided for the flange/web connection as it is possible that the flange might have to transmit a moment about the weld axis (6.7.9). As intermittent welds are to be used, the distance between the ends of adjacent welds must not exceed the lesser of:

a) 10 × 20 mm = 200 mm or 300 mm in compression: i.e. 200 mm
b) 24 × 20 mm = 480 mm or 300 mm in tension: i.e. 300 mm.

The design spacing of 200mm is thus satisfactory.

For 6063-T5 web and flange material, the filler wires and rods for inert-gas welding are found using Section 2.5.3.2 and Table 2.8 (6.8.3).

Type 4/5 will have a maximum strength below the fully heat-treated parent metal strength (Table 2.8).

Type 4 will give the maximum resistance to corrosion and freedom from persistent weld cracking (Table 2.8(1)). Note that if the factored resistance of the weld is greater than that of the adjacent material in the HAZ, the joint deformation capacity would be improved (6.8.1).

Type 4 wire is chosen in accordance with Section 2.5.3.2 and Table 2.8 giving $p_w = 150\ N/mm^2$ (6.8.3 and Table 6.2).

Appendix D shows that higher weld strengths are available (D.2.1 and Table D.1). Type 4 and Type 5 filler material groups have a BS alloy designation (Table 2.4).

If 5356 or 5056A of Type 5 is used, $p_w = 155\ N/mm^2$ (Table D.1).

If 5183 or 5556A of Type 5 is used, $p_w = 165\ N/mm^2$ (Table D.1).

If Type 4 is used, $p_w = 150\ N/mm^2$ (Tables 2.4 and D.1)

Alloy 6063-T5 has a durability rating of B (Table 2.1), while Type 5 filler has a durability rating of A and Type 4 of B (Table 2.4).

For fillet welds $(\sigma_1^2 + 3[\tau_1^2 + \tau_2^2])^{0.5} \leq 0.85 p_w/\gamma_m$ (6.9.2). $\gamma_m = 1.3$ (Table 3.3).

For a simple longitudinal fillet weld $\sigma_1 = \tau_1 = 0$, which will give

$$P_{RF} = (0.85 p_w l_f g_t)/(\gamma_m 3^{0.5}).$$

l_f is the effective length of the fillet weld, which for the weld connecting a flange and a web of a plate girder is the same as the effective length of a butt weld (6.9.2).

$l_f = l_e = 2 \times 1000\ mm = 2\ 000\ mm$ as there is a fillet weld on each side of the flange.

The throat thickness if the fillet weld is given as $g_t = 7\ mm$.

$$P_{RF} = (0.85 \times 150\ N/mm^2 \times 2\ 000\ mm \times 7\ mm)/(1.3 \times 3^{0.5})$$
$$= 792.75\ kN.$$

The weld design of $P_{RF} = 792.75\ kN$ is satisfactory for the load of 750 kN applied parallel to the length of the weld.

11.12 Design example 3

A butt weld is subjected to a normal stress perpendicular to the throat section and also to a shear load on the throat section parallel to the weld axis that produce stresses of $\sigma_1 = 50\ N/mm^2$ and $\tau_2 = 55\ N/mm^2$ (BS 8118: Pt 1, Figure 6.4). The parent metal is alloy 6063-T5 and the filler wire is Type 4. Determine if the butt weld is suitable.

Using 11.11, 'Design example 2',

$p_w = 150\ N/mm^2$ (Table 6.2) and
$\gamma_m = 1.3$ (Table 3.3).

Type 4 wire is used in accordance with Section 2.5.3.2 and Table 2.8 (6.8.3).

For the butt weld $(\sigma_1{}^2 + 3\tau_2{}^2)^{0.5} \leq p_w/\gamma_m$ (6.9.1) giving

$([50\ N/mm^2]^2 + 3[55\ N/mm^2])^{0.5} \leq 150\ N/mm^2/1.3$
$= 107.59\ N/mm^2 \leq 115.38\ N/mm^2$.

The butt weld is therefore suitable.

11.13 Design example 4

A butt weld in alloy 6063 is subjected to an external oblique tensile load P at an angle of $\theta = 45°$, where θ is the angle between the line of the butt weld and the line of action of P shown in Figure 11.1 overleaf (BS 8118: Pt 1, Figure 6.4). The effective weld length l_e is 300 mm. The effective throat thickness t_e is 18 mm and complies with Section 6.7.8 (Figure 6.1). Run-on and run-off plates will be used (6.9.1). The parent metal on both sides of the weld is of the same thickness. $p_w = 150\ N/mm^2$ (Table 6.2), $\gamma_m = 1.3$ (Table 3.3). Determine the maximum values of P.

$P \leq P_{RB} = (p_w l_e t_e [1 + 2\ Cos^2\theta]^{-0.5})/\gamma_m$ (6.9.1)
$= (150\ N/mm^2 \times 300\ mm \times 18\ mm[1 + 2\ Cos^2 45°]^{-0.5})/1.3$
$= 440.58\ kN$.
$P \leq 440.58\ kN$.

If the weld is subjected to in-plane bending, the factored resistance per unit length of weld is

$P_{RB}/l_e = 440.58\ kN/300\ mm = 1.469\ kN/mm$ length of weld (6.9.1 Note 4).
$P \leq 1.469\ kN/mm$ length of weld.

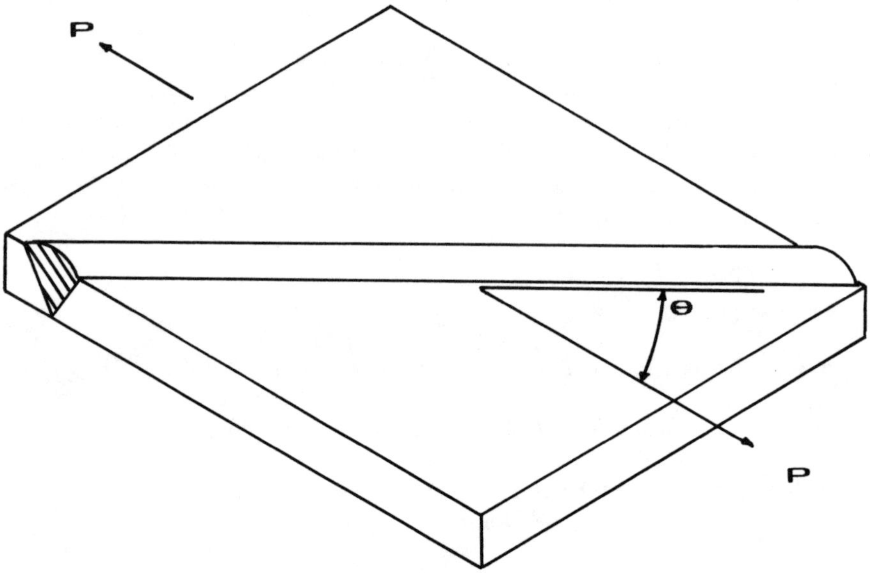

Figure 11.1 Butt weld design

If the butt weld lacks external shear loading and $\theta = 90°$, then (6.9.1)

$$
\begin{aligned}
P_{RB} &= (p_w l_e t_e)/\gamma_m \\
&= 150\ \text{N/mm}^2 \times 300\ \text{mm} \times 18\ \text{mm}/1.3 \\
&= 623.08\ \text{kN.}
\end{aligned}
$$

$P \leq 623.08$ kN.

If the load had been a shear load parallel to the line of the butt weld (6.9.1), then

$$
\begin{aligned}
P_{RB} &= (p_w l_e t_e)/(3^{0.5}\gamma_m) \\
&= 150\ \text{N/mm}^2 \times 300\ \text{mm} \times 18\ \text{mm}/(3^{0.5} \times 1.3) \\
&= 359.73\ \text{kN.}
\end{aligned}
$$

$P \leq 359.73$ kN.

11.14 Design example 5

A HSFG bolt conforming to BS 4395 is to be used to secure two 15 mm thick plates in alloy 6082-T6. The plates will not be subjected to extremes of temperature and the bolt grip length is short (6.5.1). The bolt is 50 mm from the bearing edge (6.3.3). The bolt diameter is 24 mm, $A_{es} = A_{tb} = 353\ mm^2$, $P_o = 207$ kN and $p_f = 627\ N/mm^2$. Normal clearance is assumed (6.4.2). The friction interfaces have been treated to comply with Section 6.5.6 of BS 8118: Pt 1 and $\mu_s = 0.33$. The bolt is not subjected to any axial applied external tensile load S_{tb}. Determine the joint factored resistance and the serviceability limit state of deformation.

The bolt material will be steel or corrosion-resisting steel in accordance with BS 4395. The thermal expansion of the aluminium connected parts will have no significant effect (6.5.1). The joint design requires the alloy of the connected parts to have a proof strength exceeding 230 N/mm^2 (6.5.1). For alloy 6082-T6 plate, the minimum 0.2 per cent proof stress is 240 N/mm^2 (Table 2.1) and is thus satisfactory.

For the ultimate limit state of static strength, the ultimate capacity is the lesser of the bearing capacity and the shear capacity (6.5.2).

For the bearing capacity, the lesser of B_{RF} and B_{RP} is used (6.4.4):

$$
\begin{aligned}
B_{RF} &= 2d_f t p_f/\gamma_m \\
&= 2 \times 24\ mm \times 15\ mm \times 627\ N/mm^2/1.2 \\
&= 376.2\ kN.
\end{aligned}
$$

B_{RP} is the lesser of $cd_f tp_a/\gamma_m$ and etp_a/γ_m.

$d_f/t = 24\ mm/15\ mm = 1.6$, therefore $c = 2$ (6.4.4). $e = 50$ mm.
p_a is the lesser of $0.5(f_{0.2} + f_u)$ and $1.2f_{0.2}$ (Table 2.1) for alloy 6082-T6 plate of the connected ply giving:

$$
\begin{aligned}
p_a &= 0.5(240\ N/mm^2 + 295\ N/mm^2) \\
&= 267.5\ N/mm^2 \text{ or} \\
p_a &= 1.2 \times 240\ N/mm^2 \\
&= 288.0\ N/mm^2 \text{ giving} \\
p_a &= 267.5\ N/mm^2.
\end{aligned}
$$

$$
\begin{aligned}
B_{RP} &= 2 \times 24\ mm \times 15\ mm \times 267.5\ N/mm^2/1.2 \\
&= 160.5\ kN \text{ or} \\
B_{RP} &= 50\ mm \times 15\ mm \times 267.5\ N/mm^2/1.2 \\
&= 167.19\ kN.
\end{aligned}
$$

Thus $B_{RP} = 160.5$ kN is the limiting value as it is less than $B_{RF} = 376.20$ kN.

For the shear capacity (6.4.2):

$$V_{RS} = \alpha_s p_f A_{es} K_1/\gamma_m$$

$= 0.7 \times 627\ N/mm^2 \times 353\ mm^2 \times 0.85/1.2$
$= 109.74$ kN.

The ultimate capacity is the lower of $V_{RS} = 109.74$ kN and $B_{RP} = 160.5$ kN and is 109.74 kN.

The serviceability limit state of deformation is reached when the bolt shear load equals its friction capacity (6.5.3).

For the friction capacity F_c (6.5.4):

$$F_c = P_p \mu_s N_f / \gamma_s$$

where:
$P_p = P_o - 0.9S_{tb} = (207\ kN - 0.9 \times 0.0)$ as
$P_o = 207$ kN (BS 4395 and 6.5.5) and S_{tb} is zero.
$\mu_s = 0.33$ as 2t (2×15 mm) $> d_f$ (24 mm) and the gross area stress will be limited to $< 0.6f_{0.2}$ ($0.6 \times 240\ N/mm^2 = 144\ N/mm^2$), $N_f = 1$ and $\gamma_s = 1.33$ giving

$$F_c = 207\ kN \times 0.33 \times 1.0/1.33 = 51.36 kN.$$

The serviceability limit state of deformation is reached when the bolt shear load equals $F_c = 51.36$ kN (6.5.3). But V_{RS} (109.74 kN) $> F_c$ (51.36) and the value of μ_s would need to be increased or additional HSFG bolts used.

11.15 Design example 6

A pin joint with a single pin in double shear is in tension in a permanent installation (6.6.1). The line of action of the load is in the direction of the grain flow of the two connected parts (6.6.3). The diameter of the pin d_f is 40 mm and the diameter of the hole is 41 mm. The pin is 100 mm from the bearing end. The outer forked ends are 40 mm thick and the inner bearing part is 70 mm thick. There is a gap of 5 mm either side of the inner bearing between the forked ends as shown in Figure 11.2 opposite. The pin is made from alloy 6082-T6 extrusion and the connecting parts from alloy 5083-O extrusion. $\gamma_m = 1.2$ (Table 3.3).

For the pin alloy 6083-T6 extrusion $p_a = 290\ N/mm^2$ (Table 4.1). For the pin alloy 5083-O extrusion $p_a = 150\ N/mm^2$ (Table 4.2).

All the plates (70 mm and 40 mm) have a thickness greater than half the pin diameter (40 mm/2 = 20 mm) and variations of the bearing pressure across the plate thickness may be considered (6.6.2). In this case, we will assume uniform bearing pressure to determine the effective span and take the load as acting at the centre of the bearing areas.

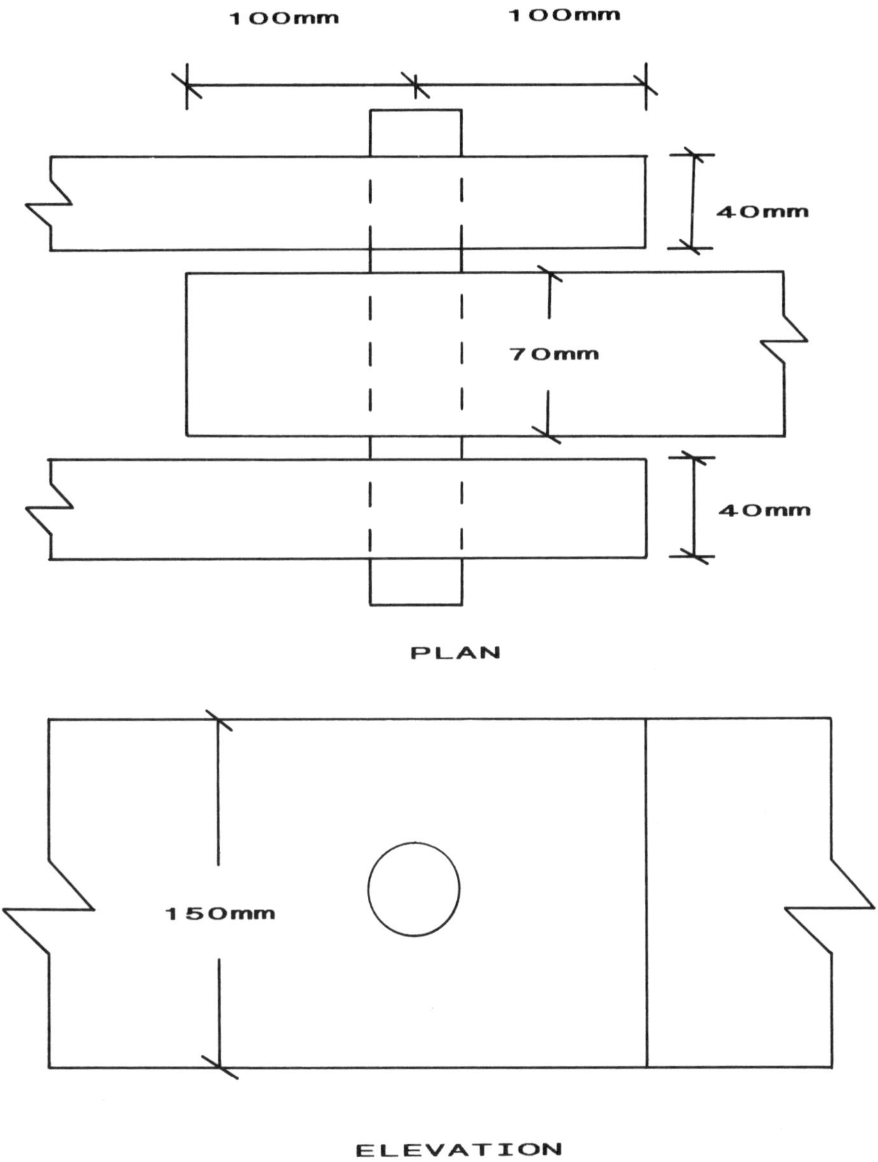

Figure 11.2 Elevation and plan of the pin joint

Under factored loading, the mean shear stress in the pin should not exceed $0.6p_a/\gamma_m$ (6.6.2).

Thus $P \leq 2 \times (\pi \times [20\text{ mm}]^2) \times 0.6 \times 290\text{ N/mm}^2/1.2$
≤ 364.42 kN.

In bending, the stress must not exceed:

$1.2p_a/\gamma_m = 1.2 \times 290\text{ N/mm}^2/1.2$
$= 290\text{ N/mm}^2$ (6.6.2).

For bending the effective span is taken as the distance between the centre of the bearings = (70 mm + 5 mm + 5 mm + 20 mm + 20 mm) = 120 mm (6.6.2).

As the pin will be in a permanent installation, then a fully plastic distribution of bending stress may be assumed. This means using S, the plastic modulus, where S = Z times the shape factor = $(4r^3/3)$ = 10 667 mm^3 (6.6.2).

The maximum value of the bending moment M = S times the bending stress = 10 667 $\text{mm}^3 \times 290\text{ N/mm}^2$ = 3.093 kN-m.

Assuming point loading on the pin, M = WL/4, giving

W = 4M/L = 4 × 3.093 kN-m/120 mm = 103.1kN.

The maximum factored load in shear = 364.42 kN. The maximum factored load in bending = 103.1 kN giving P = 103.1 kN as the maximum load on the pin.

The net area across the pin hole, normal to the axis of the pin-connected tension member, should be at least (6.6.3) $1.33P\gamma_m/p_a$, where P is 103.1 kN and $p_a = 150\text{ N/mm}^2$ for the member, giving the required area as $1.33 \times 103.1\text{ kN} \times 1.2/150\text{ N/mm}^2 = 1\,097\text{ mm}^2$.

For the centre tension member, the net area across the pin hole is 70 mm x 40 mm = 2 800 mm^2, which is larger than 1 097 mm^2, therefore satisfactory (6.6.3).

The thickness of the connected member should be at least $P\gamma_m/(1.6p_a d_f) = 103.1\text{ kN} \times 1.2/(1.6 \times 150\text{ N/mm}^2 \times 40\text{ mm}) = 12.89$ mm, which is less than the actual thickness of 70 mm or 2 × 40 mm, therefore satisfactory.

The net area of any section on either side of the axis of the member, measured at an angle of 45° or less to the axis of the member, should be at least $0.9P\gamma_m/p_a = 0.9 \times 103.1\text{ kN} \times 1.2/150\text{ N/mm}^2 = 742\text{ mm}^2$ (6.6.3). For the centre member, the net area is 70 mm × 150 mm = 10 500 mm^2, therefore satisfactory.

The bearing plate net width is 150 mm ≤ (8 × 70 mm) and 150 mm ≤ (8 × 40 mm), therefore satisfactory (6.6.3).

The pin hole diameter of 41 mm < 105 per cent × 40 mm (42 mm), therefore satisfactory (6.6.3). The pin retaining system must be able to withstand a lateral load of P/10 = 103.1 kN/10 = 10.31 kN.

The limitation on the loading is related to the bending strength of the pin. A reduction in the loaded span would increase the allowable value of P. P would also increase if the actual bearing area and the pressure distribution on the pin were used.

11.16 Design example 7

A tie member 350 mm wide, 25 mm thick, in alloy 5083-O plate material, is in axial tension. The member contains five holes for 22 mm diameter steel bolts. The bolts are not close-fitting and have $p_f = 240\ N/mm^2$. The fastener is not subjected to reversal of load. All edges are machined and the outside plies are 25 mm thick. Figure 11.3 overleaf gives the proposed bolt locations.

a) Determine the bolt spacings.

The joint will not be a long joint as the joint length will be less than 330 mm ($15d_f$) where $d_f = 22$ mm. The number of fasteners in the direction of the transmission of the load will not exceed five (6.2.4). The minimum spacing is 2.5 × 22 mm = 55 mm (6.3.1).

For tension members, the maximum spacing on a line in the direction of the stress is the lesser of 200 mm or 400 mm (16 × 25 mm). In compression or shear, the maximum spacing is 200 mm or 200 mm (8 × 25 mm). The spacing of the bolts on a line adjacent and parallel to an edge on the outside ply must not exceed 100 mm or 200 mm (8 × 25 mm). If the bolts are staggered on adjacent lines not more than 75 mm apart, the 200 mm may be increased to 300 mm and the 100 mm increased to 150 mm (6.3.2). In any event, for the 25 mm thick plate, the maximum tension, compression or shear spacing is 300 mm.

The minimum edge distance is 33 mm (1.5 × 22 mm) and on the bearing side 44 mm (2 × 22 mm) (6.3.3).

The hole clearance on the diameter is ≤ 0.8 mm giving a hole diameter of 22.8 mm (Table 3.1, BS 8118: Pt 2) (6.3.4). There is no packing or countersinking (6.3.5 and 6.3.6).

b) Determine the factored resistance of each bolt in single shear, axial tension and in bearing.

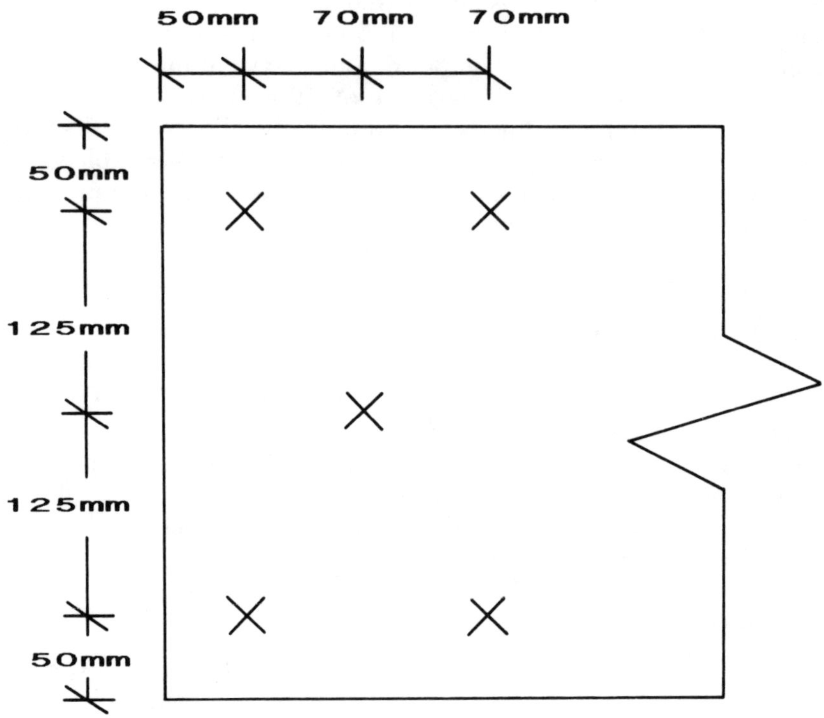

Figure 11.3 Proposed bolt locations

For shear:

$V_{RS} = \alpha_s p_f A_{es} K_1/\gamma_m$

where:
α_s = 0.7,
p_f = 240 N/mm^2,
$A_{es} = A_{tb}$ = 303 mm^2,
K_1 = 0.85 and
γ_m = 1.2 (6.4.2) giving

$V_{RS} = (0.7 \times 240 \text{ N/mm}^2 \times 303 \text{ mm}^2 \times 0.85)/1.2$
$= 36.06$ kN.

For axial tension:

$P_{RT} = ap_f A_{tb}/\gamma_m$

where:
a $= 1.0$ and
$A_{tb} = 303\ mm^2$ (6.4.3) giving

$P_{RT} = 1.0 \times 240\ N/mm^2 \times 303\ mm^2/1.2$
$= 60.6\ kN$.

For bearing:
The effective factored resistance in bearing is the lesser of B_{RF} and B_{RP} (6.4.4).

$B_{RF} = 2d_f t p_f/\gamma_m$

where:
$d_f = 22$ mm and
t $= 25$ mm giving

$B_{RF} = 2 \times 22\ mm \times 25\ mm \times 240\ N/mm^2/1.2$
$= 220.0\ kN$.

B_{RB} is the lesser of $cd_f t p_a/\gamma_m$ or etp_a/γ_m

where:
$c = 2$ as 22 mm/25 mm < 10 and
$e = 50$ mm (Figure 11.3).

For alloy 5083-O plate, p_a is the lesser of $0.5(f_{0.2} + f_u) = 0.5(125\ N/mm^2 + 275\ N/mm^2) = 200\ N/mm^2$ and $1.2f_{0.2} = 1.2x125\ N/mm^2$ and is given as $p_a = 150\ N/mm^2$ in Table 4.2.

$B_{RP} = 2 \times 22\ mm \times 25\ mm \times 150\ N/mm^2/1.2$
$= 137.5\ kN$ or
$B_{RP} = 50\ mm \times 25\ mm \times 150\ N/mm^2/1.2$
$= 156.25\ kN$ giving
$B_{RP} = 137.5\ kN$.

In bearing the limit is $B_{RP} = 137.5\ kN$.

c) Determine the factored tension resistance P_{RS} (4.6.2).

P_{RS} is the lesser of: for general yielding (4.6.2.2) $P_{RS} = p_o A/\gamma_m$ and for local failure (4.6.2.3) $P_{RS} = p_a A_n/\gamma_m$.

For alloy 5083-O:

$p_o = 105\ N/mm^2$,
$p_a = 150\ N/mm^2$ (Table 4.2) and
$\gamma_m = 1.2$ (Table 3.3)

For general yielding, end connections are ignored and there are no HAZ effects:

$$P_{RS} = (105\ N/mm^2 \times 350\ mm \times 25\ mm)/1.2 = 765.63\ kN.$$

For local failure and again no HAZ effects:

$P_{RS} = (150\ N/mm^2 \times A_n)/1.2$, but A_n must consider the staggered holes as shown in Figure 11.4 opposite (4.6.2.4).

$A = 350\ mm \times 25\ mm = 8\ 750\ mm^2$ and
$A_n = A - (\Sigma A_h - \Sigma[x^2 t/4y])$

Consider line A–B–F–G,

$$A_n = 8\ 750\ mm^2 - (2 \times 25\ mm \times 22.8\ mm) = 7\ 610\ mm^2.$$

Consider line A–B–C–D–E,

$$A_n = 8\ 750\ mm^2 - (3 \times 22.8\ mm \times 25\ mm - [\{2 \times 70^2\ mm \times 25\ mm\}/\{4 \times 125\ mm\}]) = 7\ 530\ mm^2.$$

Consider line A–B–C–F–G,

$$A_n = 8\ 750\ mm^2 - (3 \times 22.8\ mm \times 25\ mm - [\{2 \times 70^2\ mm \times 25\ mm\}/\{4 \times 125\ mm\}]) = 7\ 530\ mm^2.$$

The lowest value of A_n is 7 530 mm^2 giving

$$P_{RS} = 150\ N/mm^2 \times 7\ 530\ mm^2/1.2$$

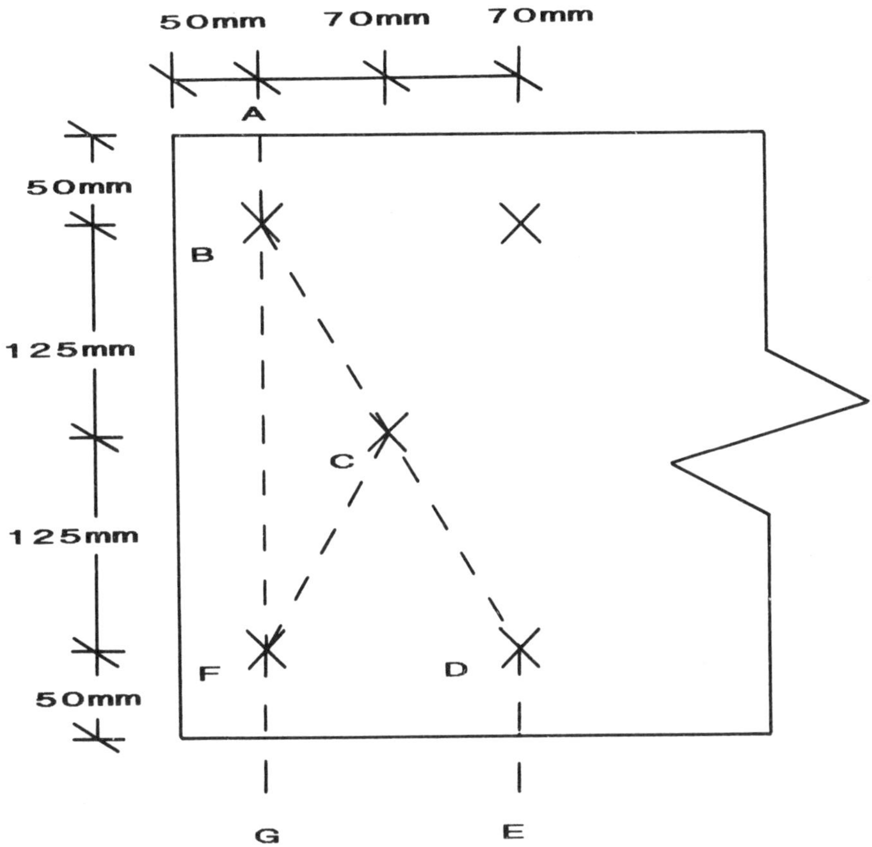

Figure 11.4 Staggered hole calculations

= 941.25 kN.

The lowest value of P_{RS} is 765.63 kN for general yielding.

The value of P_{RS} = 765.63 kN should be compared with 5 × V_{RS} = 5 × 36.06 kN = 180.3 kN and 5 × B_{RP} = 5 × 137.5 kN = 687.5 kN. Thus for axial tension, 5 × V_{RS} = 180.3 kN is the limiting value for five bolts in single shear. This value would be increased if the bolts were in double shear.

12 Fatigue (7)

12.1 General

Section 7 of BS 8118, on fatigue, is used for the fatigue calculations related to elements formed from extrusions, plates, sheet and strip. Test data may also be used. However, the fatigue conditions for castings and forgings must be obtained from the manufacturers (7.1.1).

The static limit state criteria given in Sections 3 and 4 may not be an accurate guide to the risk of fatigue failure, as fatigue failure is related to fluctuating service loads. The nominal or unfactored loading is used for checking the limit state of fatigue and vibration, while factored loads are used for checking the limit state of static strength. It is necessary to determine the extent to which fatigue may control design, and the following must be considered:

1) an accurate prediction of the full service life loading sequence,
2) the accurate assessment of the elastic response of the structure due to the full service life loading sequence and
3) a more precise definition of the degree of quality control, detail design and the methods of manufacture (7.2.1).

12.2 Mechanism of failure (7.1.3)

The mechanism of failure in aluminium usually includes the following:

1) fatigue failure initiation at a point of high stress concentration,
2) fatigue cracks extending incremental due to cyclic stress change,
3) fatigue cracks remain stable under constant load and
4) ultimate failure occurs when the remaining cross-section cannot sustain the applied peak tensile load.

Further:

1) fatigue cracks propagate at right angles to the direction of the maximum principal stress range,
2) the rate of crack propagation is proportional to: at least ([stress range]^3 and the [total crack length]$^{0.5}$) and
3) the early stages of crack growth are slow, often inconspicuous and give rise to problems of in-service detection (7.1.3).

The most usual sites for fatigue crack initiation are at (7.1.4):

1) machined corners and drilled holes,
2) the roots of fastener threads,
3) toes and roots of fusion welds and
4) surfaces under high contact pressures (fretting).

The main conditions that affect fatigue performance are (7.1.5):

1) Application of loads (obviously!): A high number of load cycles affects the design life. Members with low natural frequencies, or slender structures, even when the static design stresses are low, are prone to resonance and the magnification of dynamic stress. This is especially the case for fluid loading, including wind, and for structures supporting machinery.
2) Environment: Certain types of chemical and thermal environments reduce fatigue strength.
3) High ratio of dynamic loading to static loading: Lifting and/or moving structures, such as cranes, land transport vehicles and sea transport vehicles, etc., are more likely to have fatigue problems than fixed structures. Fixed structures such as bridges, which carry moving loads, also suffer from fatigue.
4) Joint detail: Complex joints cause local stiffness variations in the load path and result in high stress concentrations. Although the ultimate static capacity of the joint may be unaffected, the fatigue resistance can be severely reduced. If fatigue resistance is of predominant importance, the joint design should be simple and ensure smoothness of load transfer.

5) Welding: Many weld details reduce fatigue strength and this applies to attachments to loaded members and joints between members, whether they are structural or not.

12.3 Fatigue design and failure criteria (7.2)

Aluminium structures should be designed to provide a safe life; that is, the probability of fatigue failure during the life of the structure is comparable with that for the other ultimate limit state modes of failure (7.2.1). In economic terms, the use of the damage-tolerant approach or the fail-safe approach may be justified when related to a) the degree of redundancy, b) the ease of inspection and repair and c) the severity of loading. For this approach, the safety margins may be reduced below those required for a safe life design as given in Section 3 (7.2.1).

The required life will be achieved if:

$$f_r^m N \leq K_2$$

but structural details do not normally undergo constant amplitude stress histories. Thus Section 7.3 is used to consider general loading. f_r, the principal stress range, is found using Sections 7.4 and 7.6. Section 7.7 classifies the commonly used detail types. The values of K_2 and m are given in Section 7.8. γ_f is taken as 1.0 if the fatigue strength data in Section 7.8 and the loading complies with Section 7.4.

12.4 Fatigue assessment procedure (7.3)

Any structural member can contain many fatigue crack initiation sites. The sites containing the highest stress fluctuations and/or the severest stress concentrations are checked first, using the following procedure:

1) determine an upper bound estimate of the service loading sequence for the structure's design life (7.4 and Appendix C), then
2) estimate the resulting stress history for the detail being checked (7.5),
3) reduce the stress history to an equivalent number of cycles (n) of a different stress range (f_r), using the cycle technique of Section 7.6.1,
4) rank the cycles in descending order of amplitude, f_{r1}, f_{r2}, etc., to form a stress spectrum (7.6.2),
5) then classify the detail to Tables 7.1, 7.2 and 7.3, plus Section 7.7. Using Section 7.8.1 find the permissible endurance (N_1, N_2, etc.) for

the appropriate classification and design stress range (f_{r1}, f_{r2}, etc.). If $\gamma_{mf} \neq 1$, then the actual value of γ_{mf} must be used in setting the values of the design stress ranges (3.6.2),

6) Miner's summation is used to determine the total damage for all cycles:

The estimated life = ([the factored design life]/[Σ\{n/N\}]).

When $\Sigma(n/N) > 1.0$, either the detail should be changed to a higher class (7.7) or the stress ranges at that point should be reduced (7.3).

12.5 Fatigue loading (7.4)

All sources of fluctuating load, including the following, must be identified:

1) acceleration forces in moving structures,
2) environmental loads, including waves, wind, etc.,
3) superimposed moving loads, including machinery vibrations in stationary structures and
4) temperature changes.

Fatigue loading is described in terms of a design load spectrum and defines:

1) a range of intensities of a specific live load event and the phasing between two or more of such independent live load events plus
2) the number of times that each intensity level is applied during the structure's design life.

Specific fatigue load assessment guidance may be obtained from:

1) BS 2573 for cranes,
2) BS 5400: Pt 10 for highway bridges and railway bridges and
3) BS 8100 for lattice towers.

A realistic assessment of the fatigue loading must be made. If no relevant data exists, then data from existing structures subjected to similar loading may be used. Continuous displacement measurements and continuous strain measurements will be needed for the loading data to be determined from the response analysis. The possibility of resonance needs careful consideration (8.4.2).

The design load spectrum is treated as an upper bound estimate of the accumulated service condition over the structure's full design life. Foreseeable operational usage and environmental effects must be fully accounted for. The confidence limit on the design load spectrum is the mean, less two standard deviation limits on both the amplitude and the frequency (7.4).

12.6 Stresses (7.5)

Elastic theory is used to model the structure, when the stress response is calculated from specific load events. HAZ effects and local buckling effects should not be used to reduce the section properties; however, secondary bending stresses due to joint fixity in lattice structures must be considered as they may produce non-linear effects. Plastic redistribution of stresses is not allowed. The accurate modelling of the elastic stiffnesses of members and joints is required and should include any permanent non-structural material (7.5.1).

Where strain measurements on a similar structure are used to determine stress response data, the strain measurement transducers must be sited to record the correct stress parameters (7.5.1). Guidance on the measurement of strain data is given in Section 8.4.

For fatigue assessment, crack initiation sites and crack propagation paths must be considered as follows (7.5.2):

1) For parent metal and full penetration butt welds:

The nominal principal stress range in the member is used for cracks initiating from fastener holes, faying surfaces, weld toes, etc. that propagate through the parent metal or through the fully penetrated weld metal. Local stress concentration effects due to bolt holes, rivet holes, weld profiles, etc. are ignored as they are considered in the $f_r - N$ detail class strength data (Tables 7.1, 7.2 and 7.3). Residual stresses in welded joints may also be ignored as they are taken into account in the $f_r - N$ data. The fineness of the finite element mesh at a joint, if that method of analysis is used, must ensure that local stresses are not used (7.5.2 and 8.4.4.1).

Figure 7.2 shows a sample of the larger geometrical effects that may cause a non-linear stress distribution. The examples include:

a) changes in direction or alignment beyond those allowed in Tables 7.1, 7.2 and 7.3,
b) distortion, shear lag and warping in wide-plated or hollow members,
c) gross changes in cross-section shape, as at a cut-out,
d) gross changes in cross-section stiffness, as at angled junctions between thin wall members,
e) non-linear out-of-plane bending effects in slender components where the static stress is close to the elastic critical stress, as in the tension field in a web, and
f) secondary bending stresses as in joint fixity in lattice structures.

In mechanical joints, where the tensile residual stress has been allowed for, the part of the stress range in overall compression may be reduced by 40 per cent.

2) Fillet and partial penetration butt welds:

The vector sum of the shear stresses in the weld metal, based on the effective throat dimension shown in Figure 7.3, is used for cracks that initiate from the weld root and propagate through the weld throat.

The stress per unit length of the weld in a lapped joint is based on Figure 7.4, and is the average area for the axial forces and an elastic polar modulus of the weld group for in-plane moments.

The effect of different axial stiffnesses along a tee-joint must be considered.

The stresses at the root of incompletely penetrated butt welds and single fillet welds, subjected to out-of-plane bending, are calculated using, through the throat, a linear stress distribution shown in Figure 7.5.

In partially penetrated welded joints, no allowance is made for the bearing contact on the root face.

3) Threaded fasteners under axial load:

The mean axial stress on the core area of the thread is used when assessing cracks initiating at the thread roots. If there is also bending, the peak stress, based on the elastic modulus of the core, is used (7.5.2).

12.7 Derivation of stress spectra (7.6)

The actual stress history is broken down into a convenient spectrum of cycles, in terms of f_r the amplitude and n the frequency, called cycle counting. The reservoir method of cycle counting, shown in Figure 7.6, is used for short stress histories with repeated simple loading events. The rain flow method of cycle counting is used where there are long stress histories, such as measured strains in actual structures. Both the reservoir method of cycle counting and the rain flow method of cycle counting have the advantage of being suitable for computer analysis (7.6.1).

A stress spectrum is obtained by listing the cycles in descending order of amplitude f_r. The stress spectrum can be simplified into fewer bands, with the bands being grouped together into larger groups with the same total number of cycles, with f_r being equal to the highest band in the group, shown in Figure 7.7. The weighed average of the bands in one group is calculated using the power m, the inverse slope of the $f_r - N$ curve most likely to be used (7.6.2). The use of the arithmetic mean is not advised.

12.8 Classification of details (7.7)

For a detail, the fatigue strength is dependent upon the following:

1) the direction of the fluctuating stress relative to the detail,
2) the geometrical arrangement and relative proportion of the detail and
3) the location of the initiating crack in the detail.

The fatigue strength may also be dependent upon:

1) the degree of inspection after fabrication,
2) the method of fabrication,
3) the material, unless welded, and
4) the product form.

The more commonly used details are classified as follows:

1) Table 7.1: type 1, non-welded details,
2) Table 7.2: type 2, welded details on the surface of a loaded member and
3) Table 7.3: type 3, welded details at end connections.

The weld detail being used is related to the most similar detail in the tables, then the class for the particular crack initiation site associated with the detail is checked in the relevant table. For details not fully covered in Tables 7.1, 7.2 and 7.3, either published data should be consulted, or fatigue acceptance tests carried out to Section 8.4.4.1 (7.8.2).

12.9 Fatigue strength data (7.8)

Figure 7.8 shows the general form of the $f_r - N$ relationship. The design curve represents the mean, minus two standard deviation levels below the mean line through the experimental data (7.8.1).

The value of f_{oc} is set at 10^7 cycles. Below 10^7 cycles, the constant amplitude stress cycles are assumed to be non-damaging. Occasional cycles above 10^7 will cause lower amplitude cycles to become damaging and the slope of the $f_r - N$ curve of Figure 7.8 is changed to $(1/[m + 2])$ between 5×10^6 and 10^8 cycles for general spectrum loading conditions (7.8.1). Stress cycles below f_{ov} which occurs at 10^8 cycles are assumed to be non-damaging.

The use of $(1/[m + 2])$ is conservative for some spectra and it may be more appropriate to use component testing (8.4.4.1) or to use fracture mechanics analysis. Figure 7.9 gives the $f_r - N$ curves for variable amplitude stress histories, with Figure 7.8 giving a typical $f_r - N$ relationship where $N = K_o/f_r^m$, the mean line, and $N = K_2/f_r^m$, the design line. Table 7.4 gives the values of K_2 and m used in Figure 7.9 (7.8.1). Appendix L gives guidance on the derivation of the $f_r - N$ data and when conditions might allow higher strengths (7.8.2).

The static design stress for a detail must not be exceeded by the maximum tensile stress of the design stress range as found from Figure 7.9. This problem may occur in the endurance range 10^3 to 10^5 and is noted in Figure 7.9 (7.8.3).

Many detail types shown in Figures 7.1, 7.2 and 7.3 can, in certain circumstances, have their fatigue strength improved by certain manufacturing techniques. These techniques are usually used to overcome existing design deficiencies. They are expensive and present quality control difficulties that require specialist advice. The following manufacturing techniques are most effective for high cycle applications (7.8.4):

1) at the location of crack initiation, compressive residual stresses may be introduced by peening at a transverse weld toe and by the cold expansion method at bolt holes and
2) at the location of crack initiation, the stress concentration may be reduced, for example, by grinding transverse weld toes to a smooth profile.

In Tables 7.1, 7.2 and 7.3 the maximum permitted class represents the maximum fatigue strength for each detail, unless a higher value is obtained by test. If the fatigue stressing of the detail is well below that permitted, then the required class is less than the maximum allowable class. This will bring economies to inspection, manufacturing and workmanship.

The required class at the detail is found using the lowest $f_r - N$ curve of Figure 7.9, where Miner's summation is less than 1.0 as described in Section 7.3f. A series of classes may be required at a detail, one for each direction of stress fluctuation.

To concentrate the inspections on areas that are fatigue critical, the following actions are usual:

1) the use of calculations to identify regions where class 20 is exceeded, then in those regions,
2) identify on the drawings the direction of stress fluctuations and the required class as shown in Figure 7.10, together with a general note on the drawing as described in Section 7.8.5c.

12.10 Design example 1

Over the typical 120 years design life of a highway bridge (Table C.1), a connection detail is to be subjected to the stress levels and load cycles shown in Table 12.1 opposite. The connection is a welded detail on the surface of the member. The location of potential crack initiation is at a long welded attachment in the direction of f_r away from the weld end. The manufacturing requirements are that any undercuts on member edges are to be ground smooth and that stop-starts are free of lack of fusion. The special inspection requirements are for welded surfaces to be proved free of cracks and lack of fusion by penetrant dye testing.

Determine if the estimated life is ≥ the factored design life (7.3). The fatigue strength data in Section 7.8, and the loading, complies with Section 7.4 giving $\gamma_f = 1.0$ (7.2.2).

The bridge is to be designed based on having a safe life where the probability of failure by fatigue is comparable with other ultimate limit state modes of failure (7.2.1). The nominal loads are used to check the limit state of fatigue, thus unfactored loads are used (3.2.2 and 3.6.1).

The factored design life is the nominal design life which may have been increased by a fatigue life factor γ_L (3.6.2). Further, the design stress range may have been reduced by a fatigue material factor γ_{mf} (3.6.2).

The estimated life = the factored life/($\Sigma[n/N]$)

Table 12.1 Design load spectrum for the highway bridge

Sequence No	f_{max} N/mm^2	f_{min} N/mm^2	f_r N/mm^2	n	N	(n/N)
1	200	0	200	100	7000	0.0143
2	180	80	100	3000	6.70×10^4	0.0448
3	120	40	80	10000	1.30×10^5	0.0769
4	80	20	60	30000	3.50×10^5	0.0857
5	70	20	50	60000	6.30×10^5	0.0952
6	60	20	40	90000	1.30×10^6	0.0692
7	50	20	30	4×10^5	3.25×10^6	0.1231
8	40	20	20	5×10^6	2.20×10^7	0.2273
9	30	0	30	8×10^5	3.25×10^6	0.2461
10	20	−20	40	10000	1.30×10^6	0.0077
11	10	0	10	1×10^6	infinity	0.0000
(Σ[n/N]) =						0.9903
1	2	3	4	5	6	7

where $\Sigma(n/N)$ is defined in Figure 7.1f and $(f_r{}^m N) \leq K_2$ (7.2.2). The values of K_2 and of m are given in Table 7.4. The values of f_r and of n are specified by the designer. The value of N is found from Figure 7.9 using f_r and the class number found from Tables 7.1, 7.2 or 7.3.

The estimated life is found as follows (7.3):

a) Obtain an upper bound estimate of the service loading sequence for the structure's design life (7.3a).

The unfactored design load spectrum given in Table 12.1 is an upper bound estimate of the accumulated service condition over the full design life of the structure and of the detail (3.6.1). The confidence limit on the design life spectrum is based on the mean minus two standard deviation limits on both amplitude f_r and frequency n (7.4).

I) It is known that increased crack growth will not take place during the later stages of the life of the detail (3.6.2a).

II) The assumed loading spectrum is accurate (3.6.2b).

III) Loading records will be kept during the life of the detail (3.6.2c).

IV) No change of use will take place during the structure's life (3.6.2d).

V) The nominal design life will not be increased as $\gamma_L = 1$ will be used (3.6.2).

$\gamma_{mf} = 1$ will be used (3.6.2) as:

i) load paths alternative to the detail under consideration do exist (3.6.2(2)) and
ii) the detail is not in a hostile environment (3.6.2(1)).

b) Estimate the resulting stress history at the detail being checked (7.3b).

For a highway bridge, the range and frequency of loading can be obtained from traffic survey data on similar structures on similarly trafficked roads. Estimates of future traffic loading are also available. Elastic theory is used to model the structure and determine the stresses (7.5.1). The requirements of Section 7.5.2 are followed.

c) Reduce the stress history to an equivalent number of cycles, n, of different stress ranges, f_r, using a cycle counting technique (7.3c) as described in Section 7.6.1.

It is assumed that the data given in Table 12.1 was obtained using a technique described in Section 7.6.1. For highway bridges, design calculations, backed up by on-site measurements, will give the amplitude f_r. The traffic survey data will give the value of n and computer analysis will be used to determine the phasing when two or more independent live load events (vehicles) occur (7.4). Column five of Table 12.1 gives the values of n for each sequence, 1 to 11 inclusive.

d) Rank the cycles in descending order of f_r to form a stress spectrum (7.3d) in accordance with Section 7.6.2.

Column four of Table 12.1 lists the stress ranges in descending order of amplitude f_r. Columns two and three give the maximum and minimum stresses respectively of the loading event. For fatigue analysis it is the principal stress range ($f_r = f_{max} - f_{min}$) that is required. The spectrum given in column four has been simplified into eleven bands. Each band contains the same total number of cycles, but uses the amplitude equal to that of the highest value in the band (7.6.2). Column five gives the value of n for each value of f_r.

e) The detail must now be classified in accordance with Tables 7.1, 7.2, 7.3 and Section 7.7 (7.3e).

The detail we are considering is welded with a stress fluctuation in the longitudinal direction. The welded detail is on the surface of a loaded, built-up member, therefore Table 7.2 will be used. The detail is a long welded attachment, in the direction of f_r away from the weld end. Any undercut on the member edges has been ground smooth. Stop-starts are free of lack of fusion. The welded surfaces have been proved free of cracks and lack of fusion by penetrant dye testing. The design stress area is the minimum transverse cross-section of the member at the location of potential crack initiation. This gives a classification of Type 2.3, with a maximum permitted class of 35. That is, the reference strength is 35 N/mm^2 at 2×10^6 cycles (Table 7.9).

f) The value of N for column 6 of Table 12.1 must be determined (7.3e).

Let us consider sequence (row) six of Table 12.1. Using Figure 7.9 the value of $f_r = 40\ N/mm^2$ is found on the vertical axis, then moving horizontally to the permitted class line of 35 and going down to the horizontal axis gives the endurance limit of $N = 1.30 \times 10^6$. As $n = 90\ 000$, the ratio of $(n/N) = 0.0692$ as shown in column seven of Table 12.1. The evaluation of N and (n/N) is carried out for each sequence (row) in Table 12.1 and the value of $(\Sigma[n/N])$ which equals 0.9903 is determined.

The estimated life = the factored design life/$(\Sigma[n/N])$
= the factored design life/0.9903.

The values of f_r and n given in columns four and five respectively of Table 12.1 represent the factored design life as proposed by the designer. Using Miner's summation as given in column seven, the estimated life is greater than the factored design life and the design is thus satisfactory.

If $(\Sigma[n/N])$ had been greater than 1.0, then either the stress range would have to be reduced or the detail changed to a higher class (7.3).

In Figure 7.9, for $f_r = 200\ N/mm^2$, the static strength requirements may limit f_r, and this implies that this stress may need to be checked (7.8.3). In this design the value of $f_r = 200\ N/mm^2$ has been included to represent an illegal overload. This has been known to happen to some highway bridges.

13 Testing

13.1 Introduction

Section 8, 'Testing', of BS 8118: Pt 1 gives guidance on structural testing under both fatigue conditions and static conditions (8.1).

If a structural component or a complete structure has been designed to BS 8118: Pt 1, there is no need for verification by either fatigue testing and/or static testing. However, testing is appropriate if (8.1):

a) the component or structure has not or cannot be analysed,
b) design procedures and materials other than those specified in Sections 2 to 7 inclusive of BS 8118: Pt 1 have been used or
c) there is some problem regarding the appropriateness of the design method, material or workmanship.

The extent, location, method, number, report and witnessing of the testing must be agreed between the supplier and the purchaser. The method of testing being consistent with the service conditions, the sample or service component to be tested should accurately represent in every way possible the service component. The test restraints and loading should accurately represent the service condition, the loading or combination of loads that produce the lowest component resistance should be used (8.2).

13.2 Static tests (8.3)

Static tests are used to determine if the component or the structure can carry:

a) the unfactored loads within the serviceability limit state and
b) the factored loads within the ultimate limit state.

The loads or forces, displacements and, if required, the strains, must be accurately measured and recorded. A load deflection curve must be plotted (8.3.2).

For the test sample to meet the serviceability limit state and the ultimate limit state, the nominal loads must not produce deflections more than those allowed by the serviceability limit state. Further, there must be no visible signs of distress, such as instability or non-elastic deformation. The factored loads should not produce excessive deformation, instability or signs of imminent collapse. Fifteen minutes after the removal of all loads, the measured residual deflections should not exceed 5 per cent of the deflections under load (8.3.3). If the residual deflections do exceed the 5 per cent limit, the factored loads may be applied up to ten times. If the 5 per cent limit is still exceeded, the sample is rejected (8.3.4).

A separate test from the above is the ultimate resistance test, which is often used where large numbers of similar structures are to be built. The mode of failure and the appropriate ultimate loads are recorded. The ultimate resistance should be not less than the appropriate material factor γ_m times the factored loads (8.3.5).

13.3 Fatigue tests (8.4)

Fatigue tests are used when there is insufficient data to verify the design calculations of Section 7, 'Fatigue'. For example (8.4.1):

1) if the applied loading history or spectrum is not available or cannot be theoretically calculated,
2) if member forces or local stress fields cannot be calculated or
3) if the sample differs from the information given in Tables 7.1 to 7.3.

How the loading data is obtained will depend on the type of structure (8.4.2). For fixed structures, such as bridges, crane girders, machinery supports, etc., existing similar structures subjected to the same type of loading will be used, with due allowance being made for load changes during the design life. For other fixed structures, such as chimneys, masts and offshore topside structures where environmental loading is the main consideration, existing similar structures subjected to the same type of loading over longer working periods will be used. Environmental fatigue damage can be confined to very specific loading bands and greater precision is needed in simulating the structural properties such as damping, mass, stiffness, etc. and the aerodynamic properties such as cross-sectional geometry. The loading data should be compared with the meteorological records over a typical design life. Moving structures, such as boats, rail and road vehicles, require precise definitions of the riding surface and the effects of cargo on- and off-loading (8.4.2).

For simple members, it may be possible to test only the component containing the joint or detail to be analysed, with the component ends sufficiently far from the detail not to affect the recorded stresses (8.4.3.1). The forces and stresses in the regions of interest should be the same as in the complete real structure.

For some structures it may be impracticable to isolate the test components using simple applied forces. Prototype or production structures may have to be used. Static loads may have to be applied independently, allowing the resulting stresses for the point of interest to be combined using the individual loads' influence coefficients (8.4.3.2).

It may be necessary to verify the stress history using data acquisition equipment to record either the full stress history or to perform a cycle counting operation. The cycle counting operation can be used to predict the life from the appropriate $f_r - N$ curve. If the load history is uncertain, it may be necessary to attach permanently a cycle counting device to the in-service structure (8.4.3.3).

If force spectra or stress history data is known, the component can be tested to verify the design. The component must be correctly loaded, with strain gauges used to verify the stress fluctuations. To obtain a $f_r - N$ curve, at least eight specimens must be tested to given endurances in the range 10^3 to 10^8 cycles. Reference to BS 3518 is required to obtain the correct procedures. To allow for production variations the design curve is parallel to the mean curve, but is the lower of either a maximum of 80 per cent of the strength value or two standard deviations away from the mean curve. Damage tolerance may also need to be considered. If the design history is known and suitable variable amplitude facilities are available, the specimen may be tested under the unfactored stress history (8.4.4.1).

Full-scale testing can be carried out under actual operating conditions or in a test facility using loads up to the nominal loads (8.4.4.2). The applied loads should reproduce exactly the in-service loads. Where the service loads vary in a random manner, they can be represented by an agreed equivalent series of loads or the test loads should equal the unfactored loads. The test should continue until the specimen fractures or is unable to resist the full applied loading. The test history regarding load applications and defect development should be recorded.

There are two acceptance criteria: either damage-tolerant performance or safe-life performance (8.4.5).

a) In damage-tolerant design, acceptance depends upon how long a crack takes to grow to the point where it can be detected in-service (3.6.3). Consideration is also given to rate of crack growth, critical crack length, residual safety and repair costs.
b) In safe-life design, the predicted life must be not less than the design life or the nominal design life (3.6.2). The life-to-failure determined by test must take into account the number of tests and must not be less than the factored design life:

   $(N_a/F) \geq$ factored design life

   where F is a factor dependent upon the effective number of test results available and N_a is the log mean life-to-failure.

13.4 Reporting (8.5)

For tests performed in accordance with Section 8, 'Testing', a report/test certificate is required and will cover, at least, the following areas (8.5):

a) the purpose or objective of the tests,
b) description of the load system applied and the samples tested,
c) record of the load applications and of the measured reactions,
d) a summary of the deformations, loads and stresses at critical acceptance points,
e) a record of the ultimate load, the mode of failure and any relevant observed behaviour and
f) a statement of compliance or non-compliance with the acceptance criteria.

14 Appendices to BS 8118: Pt 1

14.1 Appendix A

14.1.1 Nomenclature of aluminium products

The British Standards give information concerning the nomenclature for wrought aluminium alloys for general engineering purposes. The nomenclature for castings is designated in accordance with BS 1490 (A1).

14.1.2 Alloy groups

The alloy designation uses the international four digits system as shown below (A.2). Also included are typical applications for the alloys.

Aluminium 99.00% minimum	1***	Used for ceilings, panels and tanks.
Aluminium alloy groups, by major alloying elements:		
Copper	2***	Used much more in aeronautics than civil engineering. Used for pipes, plates and profiles.
Manganese	3***	Used for panels and roof systems.

Silicon	4***	Not often used.
Magnesium	5***	Used in welded structures.
Magnesium and silicon	6***	Used in welded structures, bolted and riveted connections.
Zinc	7***	Used in structural applications.
Other elements	8***	
Unused series	9***	

In the 1*** group, the second digit indicates the modification in impurity limits or alloying elements. The last two of the four digits are the same as the last two digits in 99.** per cent and indicate the minimum aluminium percentage (A.2.2).

In the 2*** to 8*** groups, the second digit indicates the alloy modifications. The last two digits serve to identify the different aluminium alloys in the group (A.2.3).

A letter after the four digits indicates a national variation (A.2.4).

Aluminium alloys were introduced to increase the strength of aluminium. The main alloying elements are copper, magnesium, manganese, silicon and zinc. Other elements such as nickel and chromium are also used. Chromium increases resistance to stress corrosion. Copper increases strength, but reduces corrosion resistance, ductility and weldability. Magnesium reduces the melting point but increases the corrosion resistance to sea water. Nickel increases strength in high temperature conditions. Silicon increases ductility and strength. Zinc increases strength.

14.1.3 Temper or condition

Following the four-digit designation and a hyphen, the temper or condition of the aluminium or aluminium alloy is shown by a letter and a number or numbers (A.3.1).

For the non-heat-treatable alloys, strength is increased only by strain-hardening and decreased by heating. The temper is identified by the suffix letters and symbols as follows:

F As fabricated.
O Annealed.
H Strain-hardened.

The H is followed by two or more digits. The digits show the final degree of strain-hardening. The first digit has the following meaning:

H1 Strain-hardened only.
H2 Strain-hardened and partially annealed.
H3 Strain-hardened and stabilized.

A second digit following the H1 or H2 shows the degree of strain-hardening in ascending order of temper (A.3.2).

For heat-treatable alloys, enhanced strength is derived by one or two stages of heat treatment. The first stage is solution heat treatment, the second stage is precipitation heat treatment (ageing). Ageing occurs naturally for some alloys and the second formal heat treatment may not be needed. The heat-treatable condition is identified by the suffix letters and numbers as follows (A.3.3):

O Annealed.
F As fabricated with no formal heat treatment.
T4 Solution heat-treated and naturally aged.
T5 Cooled from an elevated temperature shaping process and then artificially aged.
T6 Solution heat-treated and then artificially aged.
T8 Solution heat-treated, cold-worked and then artificially aged.

Some alloys covered by British Standards still use the existing BS temper designation system, rather than ISO 2107 (A.5.1).

Table A1 of BS 8118: Pt 1 lists the nearest foreign equivalents of 14 of the British Standards wrought and cast alloys referred to in Tables 2.1 and 2.2 of BS 8118: Pt 1. Table 2.1 lists five alloys and 14 conditions. Table 2.2 lists nine alloys and 26 conditions.

14.2 Appendix B

14.2.1 Formal statement of safety factor format in the code for static design resistance calculations

A series of definitions are given for the following:

a) design action effect (B.3) S_d,
b) design resistance (B.5) R_d,
c) factored load (B.2) $F_d = \gamma_f F_k$,
d) nominal load (B.1) F_k,
e) nominal resistance (B.4) $R_k = R_m(1 - kv)$ and
f) structural adequacy (B.6) $R_d > S_d \gamma_c$

14.3 Appendix C

14.3.1 Typical values of design life

A design life is required for the fatigue assessment procedure of Section 7.3. Table C.1 gives typical design life values. Typical values of design life vary between 120 years for a highway bridge to ten years for scaffolding and vehicle bodies.

14.4 Appendix D

14.4.1 Derivation of material limiting stresses for use in design

The following expressions are used to obtain the limiting stresses p_a, p_o and p_v for the parent metal materials listed in Tables 4.1 and 4.2. The expressions can also be used for materials not specified in Tables 4.1 and 4.2 (D.1).

1) if $f_u \leq 1.4f_{0.2}$
   then $p_o = f_{0.2}$,
   $p_a = 0.5(f_{0.2} + f_u)$
   and $p_v = 0.6p_o$,
2) but if $f_u > 1.4f_{0.2}$
   then $p_o = 1.28f_{0.2} - 0.2f_u$,
   $p_a = 1.2f_{0.2}$
   and $p_v = 0.6p_o$.

When no guaranteed value of $f_{0.2}$ and f_u are quoted, it may be assumed the values are:

1) 80 per cent of the typical values given by the manufacturers or
2) the values for the same materials in the O condition (D.1).

If all precautions are taken to avoid cracking, the enhanced values for p_w, given in Table D.1, may be used instead of the values given in Table 6.2 (D.2.1). p_w for metals not covered by Table D.1 may be found experimentally.

The following expressions are used to obtain p_{az} and p_{vz} for HAZ material listed in Table 6.3 and can be used for materials not covered by the table (D.3.1).

1) For heat-treatable materials:

$p_{az} = k'_z p_a$ and
$p_{vz} = 0.6 p_{az}$.

2) For non-heat-treatable materials:
$p_{az} = 1.2\ f_{0.2}$ and
$p_{vz} = 0.6 p_{az}$.

If only typical values of $f_{0.2}$ are available, then a value of 80 per cent of the typical value should be used in the expression for p_{az}.

For a 7*** series material the value of k'_z is found from Appendix F.2, Tables F.1 and F.2, plus Table 4.5. A more favourable value of k'_z for 1) above may be found using F.2.3, cases 1 and 2 when either:

1) a single pass weld is laid in isolation or
2) when a stricter thermal control is exercised than is normally called for in BS 8118: Pt 2.

14.5 Appendix E

14.5.1 Elasto-plastic moment calculations

This appendix gives an alternative method to that given in Sections 4.5.2 or 5.4.2 for finding M_{RS} for semi-compact or slender beam elements. The elasto-plastic method has advantages for sections where the critical element for local buckling has a value of β nearly the same as β_o and has the outermost tension fibres further from the neutral axis than the compression fibres. For example, a critical compression flange element lying closer to the neutral axis than the extreme tension material, or where a critical web element stops some distance from the extreme compression face (E.1).

14.5.2 Elasto-plastic stress pattern

An idealized elasto-plastic stress pattern is used that assumes a sudden material yield at p_o (Figure E.1). The calculations use the effective net section and a reduced thickness to cover such items as HAZ softening. A deduction for holes is made. There is no reduction in thickness for local buckling and the more favourable assumed thickness of the section used for welded elements is not applied (4.5.2.3). The element is classified as either semi-compact or slender (4.3.3.4), with $\epsilon = (250/p_o)^{0.5}$.

The limiting compressive stress p_m, for the slender element, is:

$$p_m = k_L p_o$$

where,
k_L is the local buckling factor (4.3.4.2), but with $\epsilon = (250/p_o)^{0.5}$ and
p_o is the limiting stress for the material as given in Tables 4.1 and 4.2.

For longitudinally stiffened webs, a series of p_m values will be obtained due to the sub-panels and their individual values of k_L. The k_L values are found by taking the correct width and stress pattern for each sub-panel to determine its β value (5.4.4.1). Further, the force in compression should balance the force in tension. The extreme fibre tension must not exceed p_o, with the plasticity being allowed to spread in from that face. The maximum compression side stress is the limiting value p, where $p = p_o$ for a fully compact or semi-compact element and $p = p_m$ for a slender element.

If the compression flange is fully compact the plasticity is permitted to spread in from the compression face of the section. Further, if the compression flange is fully compact and has a tongue plate next to the compression edge of a semi-compact or slender web, the distance the plasticity may spread in from the compression face is limited (E.2):

1) for a semi-compact web, the plasticity must not extend beyond the compression edge of the web and
2) for a slender web, $p = p_m$ for the web must be satisfied.

When a section is fabricated from components of differing strength (a hybrid section), the strain distribution is linear across the whole section and implies:

1) either side of the neutral axis, the stress varies linearly in the elastic zone,
2) there will be a step-change in the stress, in the plastic zone, between materials with different values of p_o and
3) if the change of materials is at the same point as the junction between the elastic zone and the plastic zone, there will also be a step-change in the stress.

14.5.3 Calculations of moment resistance

The factored moment of resistance, M_{RS}, is:

$$M_{RS} = M_u/\gamma_m$$

where:
M_u is the ultimate moment and

γ_m is the material factor as given in Table 3.3 (E.3.1).

For slender sections the stress patterns described above are used to determine M_u (E.3.2).

For semi-compact sections:

$$M_u = M_{us} + ([\beta_o - \beta]/[\beta_o - \beta_1]) \times (M_{uf} - M_{us})$$

where:
M_{us} is the moment corresponding to the adopted stress pattern (E.2.1),
$M_{uf} = Sp_o$,
S is the plastic section modulus of the gross section, with no reduction for HAZ, holes or local buckling,
β is the value of β of the critical element,
β_o is the semi-compact limiting value of β (Table 4.3) and
β_1 is the fully compact limiting value of β (Table 4.3).

14.6 Appendix F

14.6.1 HAZ softening adjacent to welds

Section 4.4 can give pessimistic values for the extent and severity of HAZ softening. Appendix F will usually give more favourable values of k_z and z (F.1.1). The extent of the HAZ may also be found experimentally as an alternative to calculation as described in Section F.5.2 (F.5.2).

As the extent, and often the severity, of the HAZ softening increases with increased interpass temperature T_o, thermal control is exercised during fabrication by letting the metal cool down between passes. Usually the maximum interpass temperature is 80°C for 7*** series alloys and 100°C for all other alloys, but stricter controls can be exercised to reduce T_o and to reduce still further the HAZ effects (F.1.2).

If the resistance of a member is governed by p_a or p_v rather than p_o, k_z of Table 4.5 is modified and redesignated as k'_z in Tables F1 and F2. k'_z is favourable for most materials and applies for:

a) local failure in tension members (4.6.2.3),
b) local squashing in struts (4.7.7) and
c) shear in beams (4.5.3).

Table F2 gives values of k_z and k'_z for materials not found in Tables 4.5 and F1.

For welds in 7*** alloys, with tensile stresses acting transversely to the weld axis, $k_z = k'_z$ and equals the 'A' value. The 'B' value is used for all other stress conditions. More favourable values of 'A' than those given in Tables 4.5 and F.1 can be used for isolated straight single pass welds without pre-heat and for other welds with stricter thermal control (F2.3).

The distance z is often overestimated in Section 4.4.3 and the following alternatives can be used:

$$z = \alpha\eta z_o$$

where:

z_o is the basic value of z, α is the modifying factor for the extent of the HAZ to allow for elevated temperature,

η is the modifying factor for the extent of the HAZ to allow for increased heat build-up and

δ, which is equal to 1, but may be reduced as follows:

$\delta = 0.75$ for a joint that has three or more valid heat paths from it or

$\delta = (1.5 + [1.3L/z_o])/(3 + [L/z_o])$ for a straight joint of length L that is less than $5z_o$.

If both values of δ apply, the lower value is used (F.3.2). In certain circumstances, a lower value of α may be used for cases Q and R of Table 4.6. η may be read from Figure F.1 if stricter thermal control is specified than required by BS 8118: Pt 2 (F.3.4).

Artificial ageing after welding can be beneficial for the 6*** and 7*** series heat-treatable alloys as it may: a) improve the strength of the weld metal, b) raise the strength in parts of the HAZ and c) reduce the time taken for the alloy to reach the point of having stable mechanical properties (F.4).

14.7 Appendix G

14.7.1 General formulae for the torsional properties of thin-walled open sections

Appendix G shows how to determine the polar second moment of area about a shear centre (G.3), the torsional constant (G.2) and the warping factor (G.4) required in the buckling calculations involving torsion for some sections. For sections not covered by the appendix, other literature must be consulted.

14.8 Appendix H

14.8.1 Lateral torsional buckling of beams

The effective length l for lateral torsional buckling of a beam is found as follows:

1) if a beam has effective lateral restraints at its ends only, l is found from Table H.1. If the effective lateral restraints are at intervals along the beam length, then l is the length between the restraints.
2) If a cantilever beam has no intermediate restraints and has no moment applied at its tip, l is found from Table H.2 (H.1.3). If intermediate restraints are provided, or there is a moment at the tip, then l is found using 1) above (H.1.1) and 3) below (H.1.2).
3) For beams subjected to loading between points of lateral support, when for example the load is applied to a compression flange and both the flange and the load are free to deflect laterally, relative to the beam centroid, then l is increased as given in Table H.1. If the beam has effective lateral restraints at intervals, then l is 1.2 times the distance between the restraints (H.1.2).
4) Also, for beams subjected to normal loading where the compression flange is laterally unrestrained with both flanges free to rotate in plan and the bearing of the bottom flange of the supports providing the only torsional restraint, again Appendix H is used (4.5.6.6).

The general expression for λ is $\lambda = \pi(ES/M_{cr})^{0.5}$ and Appendix H gives values for M_{cr} for a doubly symmetrical uniform section and for a uniform section symmetrical about the minor axis only. A uniform moment is assumed in both cases (H.2). If the moment is non-uniform, then reference must be made to published literature.

Beams of varying section throughout their length are discussed in H.3, where it is assumed that $R_f \geq 0.2$.

14.9 Appendix J

14.9.1 Torsional buckling of struts: determination of the slenderness parameter λ

Appendix J discusses the general and more rigorous determination of λ than that given in Section 4.7.5.2b and Table 4.9. The use of Appendix J will result in laborious calculations, but may also result in economies of design.

14.10 Appendix K

14.10.1 Equations to the design curves

Table K.1 gives the design curve formulae that may be used instead of the design curves of Sections 4 and 5. The design curves of Figures 4.9, 4.10 and 4.12 can be extended beyond $\lambda = 130$ by using the equations to the design curves given in Table K.1 or the curves of Figure K.1.

14.11 Appendix L

14.11.1 Fatigue strength data

Figure 7.9 has design $f_r - N$ curves found from constant amplitude test data generally between 10^5 and 2×10^6 cycles which covers the more commonly used 5*** and 6*** series alloys and alloy 7020.

The survival probability is at least 97.5 per cent for the classified details tested. Typically, the mean fatigue strength $f_r - N$ curve is 30 to 50 per cent higher than the design curve. The curves are considered safe for any condition of mean stress up to the tensile proof stress. If the applied mean stress is lower than the actual mean stress no relaxation of the curves is recommended as stress concentrations may still occur (L.1).

The change in slope of the design curve beyond 5×10^6 cycles takes into account the damage due to a high stress range in the spectrum. The lower slope of $m + 2$ in Figure L.1 is safe for most commonly used spectrum shapes (L.1).

Where fatigue dominates, or where minimum mass or minimum initial cost is important, specific fatigue strength test data may be required (8.4.4). If higher strengths are obtained from the tests, then the conditions for fabrication must not be varied during production (L.2).

15 British Standards referred to

It is not possible fully to list all the BSI publications that relate to the structural use of aluminium. However, Table 15.1 lists the BSI publications most often referred to in this book. Table 15.2 gives the BS numbers that relate to structural aluminium design and are in the index of the BSI catalogue.

Besides the BSI publications listed in Tables 15.1 and 15.2, BS 8118: Pt 1 also refers to additional publications not included here.

British Standards are reviewed, at certain intervals, by the relevant technical committee. Further, any ambiguity or inaccuracy in a BS is investigated by BSI. This means that British Standards are revised by the issue of either amendments or revised editions. It is essential, therefore, that the user of a BS ensures that the latest edition, with all its amendments, is being used. In this respect, BSI do offer an updating service.

Table 15.1 BSI publications most often referred to

BS number	*BS title*
BS 499	Welding terms and symbols Pt 1: Glossary for welding, brazing and thermal cutting Pt 2: Specifications for symbols for welding
BS 1161	Specification for aluminium alloy sections for structural purposes

BS 1490	Specification for aluminium and aluminium alloy ingots and castings for general engineering purposes
BS 1974	Specification for large aluminium alloy rivets (0.5 in to 1.0 in nominal diameters)
BS 2451	Specification for chilled iron shot and grit
BS 2573	Rules for the design of cranes
BS 3019	TIG welding Pt 1: Specification for TIG welding of aluminium, magnesium and their alloys
BS 3518	Methods of fatigue testing
BS 3571	MIG welding Pt 1: Specification for MIG welding of aluminium and aluminium alloys
BS 4395	Specification for high strength friction grip bolts and associated nuts and washers for structural engineering Pt 1: General grade
BS 4604	Specification for the use of high strength friction grip bolts in structural steelwork Metric series Pt 1: General grade
BS 4870	Specification for approval testing of welding procedures Pt 2: TIG or MIG welding of aluminium and its alloys
BS 4871	Specification for approval testing of welders working to approved welding procedures Pt 2: TIG or MIG welding of aluminium and its alloys
BS 5350	Methods of test for adhesives Part C5: Determination of bond strength in longitudinal shear
BS 5400	Steel, concrete and composite bridges Pt 10: Code of practice for fatigue
BS 6399	Loading for buildings Pt 1: Code of practice for dead and imposed loads
BS 8100	Lattice towers and masts
BS 8118	Structural use of aluminium Pt 1: Code of practice for design Pt 2: Specification for materials, workmanship and protection
CP 3	Code of basic data for the design of buildings Ch. V: Loading Pt 2: Wind loads
CP 118	The structural use of aluminium
PD 6484	Commentary on corrosion at bimetallic contacts and its alleviation
ISO 2107	Aluminium, magnesium and their alloys – Temper designations

Table 15.2 BS numbers related to subject

Subject	*BS number/CP number*
Aluminium and alloy	1161, 1470, 1471, 1472, 1473, 1474, 1490, 1728, 1974, 4300, 5812
Structural design	8118, CP 118
Use of aluminium	1207, 2037, 2997, 3495, 3987, 4868, 4873, 5286
Welding	2901, 3019, 3451, 3571, 4870, 4871, 4872

Index